KEEPING BEES
ALIVE

Sustainable
Beekeeping
Essentials

Other publications by Lawrence John Connor

Bee Sex Essentials
Increase Essentials
Queen Rearing Essentials

With Robert Muir *Bee-Sentials: A Field Guide*

With Steve Repasky *Swarm Essentials*

With Andrew Connor *BeeCabulary Essentials*

With Robert Muir and Randy Kim *Safe to Chew; An Anthology*

With Dewey M. Caron *Honey Bee Biology and Beekeeping*

With Tom Rinderer et al. *Asian Apiculture*

With Roger Hoopingarner *Apiculture for the 21st Century*

With Nikolaus and Gudrun Koeniger and Jamie Ellis *Mating Biology of the Honey Bee* (Apis mellifera)

With Bernard Mobus
The Varroa Handbook 'Biology and Control' (out of print)

Fiction
Haymakers

Keeping Bees Alive

Sustainable Beekeeping Essentials

LAWRENCE JOHN CONNOR

WICWAS PRESS

Wicwas Press, LLC

Copyright © Lawrence John Connor 2019

Published 2019 by
Wicwas Press LLC
1620 Miller Road, Kalamazoo, MI 49001 USA

www.wicwas.com

ISBN paperback book: 978-1-878075-58-1
ISBN ebook: 978-1-878075-58-1

Printed and bound in the United States of America
Editor: Randy Kim
Readers: Charlotte Hubbard, Dewey Caron, Ann Marie Fauvel and
Tammy Horn Potter
Cover photo credits:
Top row: Shana Way (pollen). Second row: Bo Sterk (comb in drum hive), Dorothey Morgan (queen). Fourth row: Ted Jones (tulip poplar).

Photo code:
U = Upper
L = Lower
M = Middle
ML = Middle Left
MR = Middle Right
UL, UR = Upper Left, Upper Right
LL, LR = Lower Left, Lower Right

CONTENTS

DEDICATION

I humbly dedicate this book to the men and women who were foundational in my entomology and beekeeping education and training: Roger Hoopingarner, Al Dowdy, E.C. "Bert" Martin, S.E. "Sam" McGregor, Walter Rothenbuhler, Victor Thompson, Charles Divelbiss, G.H. "Bud" Cale Jr., Harvey York, Otto Mackinson, Harry Laidlaw, Jr., Karl Showler and Dame Eva Crane.

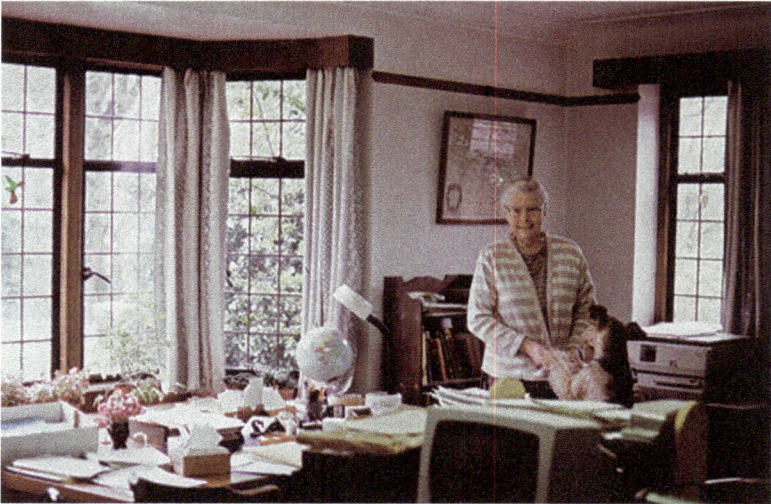

Dr. Eva Crane, MBE, founder of the International Bee Research Association in England, pulled together world beekeeping researchers to coordinate research, communication and education. She started and edited several journals and wrote comprehensive books for beekeepers and researchers.

INTRODUCTION

The worst thing a new beekeeper can do in their first year is not get any training or advice. Besides the basics, beekeeping involves a long list of critical and complex topics often overlooked during the first years of beekeeping, including bee nutrition, varroa mites, small hive beetles, American foulbrood, foraging areas, drone-laying queens, laying workers, mating failures, and the impact of pesticides on bees. True, many new apicultural clubs have formed over the past few years, but these and other topics may or may not be adequately covered at their meetings on an annual basis. The scary fact is that only a low percentage of potential, new and active beekeepers belong to a beekeeping club, read a national beekeeping magazine, attend a university extension class or do *anything* to obtain essential information to keep their bees alive.

A surprising number of beekeepers—first timers and multiyear beekeepers—are known only to the person who sells them package bees. Beekeepers often receives little or no training. This is unacceptable.

Geography poses another issue not discussed often enough. When bees are shipped from another region of the country, they usually lack the local adaptations needed to survive in the

Every beekeeper must learn how to make up new colonies. These increase nuclei or nucs may become full-sized colonies and can be used to provide additional brood, honey, bee bread, bees and a laying queen to support and boost other colonies in an operation. Think of them as insurance hives.

beekeeper's apiary. While not quite as bad as Chihuahuas running the Iditarod, you get the idea.

There is another wrinkle—now, farm supply firms and discount chain stores are selling beekeeping equipment—increasing the separation of beekeeping customers and local beekeeper wisdom. This has made the knowledge gap much, much bigger. All beekeepers, from the first timer to the person with 20 years of experience, must work hard to keep bee colonies alive and thriving. Doing it alone can be a huge mistake.

The primary goal of this book is to support and train all beekeepers—new, first year, and experienced—to sustain and grow their colony numbers. They must attempt to keep at least 85% of their bee colonies alive throughout the year[1]. All beekeepers must make sure they are capable of supporting their bees. We want these beekeepers to make sure the bees have enough forage to manage environmental factors that directly affect the growth of the colony.

We encourage every beekeeper to produce at least one increase nucleus hive from each hive they own. If possible, they should do this every season. If that expands a beekeeper's colony holdings beyond the number they want, the beekeeper can always sell their locally produced nuclei to other beekeepers in the area, reducing the need to purchase colonies from outside the region. If the queens (and the drones with which she mates) used to establish the nuclei are locally-reared, the colony will be better adapted to local conditions. Local bees require less fossil fuel than shipping queen across country.

Fire[2] is the traditional method to destroy American foulbrood, a spore-forming bacillus. Fire is final and effective but extremely wasteful if improperly used. If you have spent hundreds of dollars on your equipment, you need to get an expert's opinion to examine and test the comb before starting an inferno. Fire is not necessary for other diseases and pests. File.

The second 'worst thing' a beekeeper can do in their first year is to take bad advice. It is common for new beekeepers, or 'newbees', to lose most or all their bee colonies in their first year, and bad advice can be part of this. Maybe the queen failed as they often do, or a heavy mite load weakened the winter cluster, creating a tiny cluster unable to access stored honey just inches away during the freezing weather. Colony losses can be heavy for experienced beekeepers as well.

It would be a shame for someone to treat a problem as simple as starvation by burning all their equipment—frames, hive bodies, lids, and bottom boards—because a self-appointed online expert blamed a colony's death on American foulbrood. This unnecessary and unsustainable destruction of expensive beekeeping equipment happens way too often, and can be easily and totally avoided by some investment in proper training and education.

Another new beekeeper may not find the queen in their colony and asks for advice about what to do. Too often, the beekeeper is advised to buy and install a new queen only to later discover that the colony was simply replacing the old queen with a new one—naturally and biologically! They wasted the time and money on a new queen when they should have already been trained to know what to look for and that the bees had everything under control. Sometimes the beekeeper must know when to step back away from the bee hive and let the bees do their work uninterrupted.

Beekeeper ignorance about bee biology kills many colonies. Beekeepers must understand the many subtleties of beekeeping, from recognizing the nature of the problem to effectively determining a course of action needed to minimize colony loss. Simple management practices, like feeding new package colonies, are things newbees should understand *before* they purchased their first bees. It is also critical for returning beekeepers to master before they continue keeping bees.

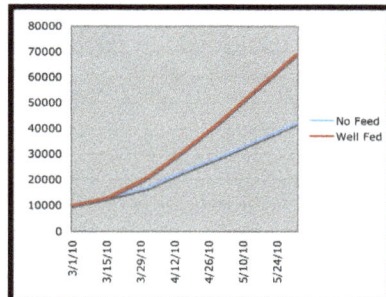

New beekeepers must learn the importance of feeding colonies during establishment and natural food shortages. This graph shows the impact of a 25% increase in egg-laying because of feeding[3].

Those of us who teach beekeeping are frustrated to find so many improperly educated beekeepers. Newbees who go to clubs after experiencing huge colony losses too often seem oblivious to the realities of the current world of bees and beekeeping. They fail to understand that the strength of their colonies—both in population size and amount

Starvation. There is no honey in the corners of the frame. The bees crawled head in to the cells, where they consumed the last of the stored honey and concentrated the last of their bodies' heat. File.

of stored food—interacts to determine the survival of a colony.

A colony needs a healthy population of young, mite-free worker bees produced by a well-raised, well-mated, and mite-tolerant queen. When newbees obtain packages of bees in the spring, they should know the type of stock the queen represents. Or they should ask. And if they didn't ask, newbees should be prepared

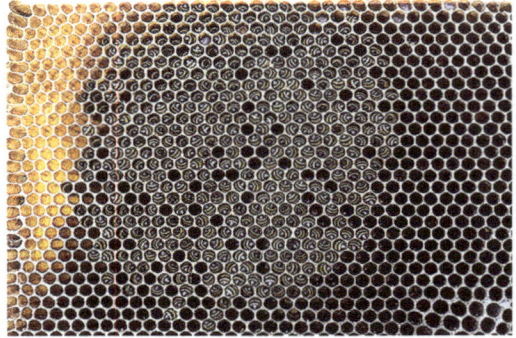

Four feeding system: UL: Jar feeds on nuclei, UR: Can feeders on cell builders, LL: Feed syrup cans inside an empty hive body, and LR: Frame feeder (on right).

Total US managed honey bee colonies Loss Estimates

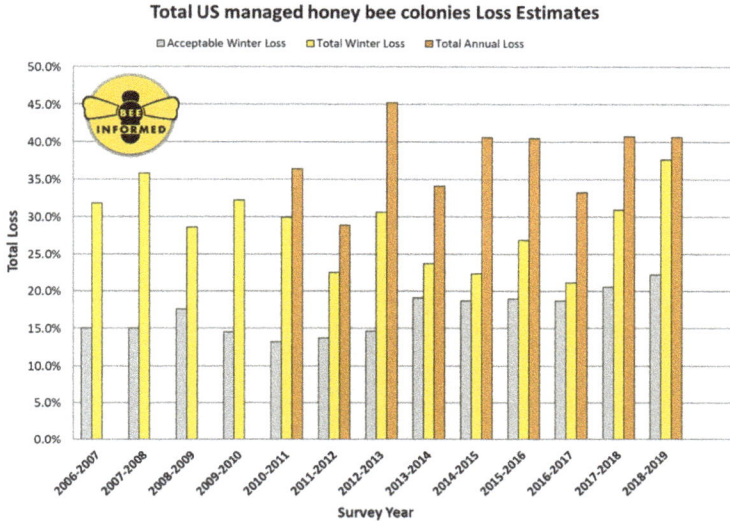

Total managed colony loss in the United States from 2006 to 2019. Bee Informed Partnership's National Honey Bee Colony Loss Survey.[4] The 2019 preliminary total losses were 40.7% with the overwinter loss of 37.7%. This was the highest in the 13 years of sampling; above the 12 year average of 28.8%. Beekeepers have accepted an increasing level of expected loss since 2010-2012.

to find their hives infested with varroa mites (*Varroa destructor*), a parasite responsible for the loss of hundreds of thousands of colonies in the United States since its 1987 introduction.

Any loss due to a lack of education is preventable with proper training. My advice? Don't unnecessarily burn up your operation.

While varroa mites appear on worker bees, the real damage is hidden with mites feeding on adult bees between abdominal sclerites and in sealed the brood of workers and drones. See photos of these elsewhere in the book. Bauer, USDA.

What is an Acceptable Level of Loss?

Colony losses are huge in the world of beekeeping, even for an experienced beekeeper, reaching almost an average of 41% in 2018-2019. Losses occur at any time for any number of reasons. It is not always about parasites and starvation, as no single factor has been identified. Maybe it's the cost of living our complicated, over-populated, human-centric and highly polluted world.

Take my friend Jimmy for example. A native of south Texas, he keeps bees on ranches, some covering 10 to 25 thousand acres. He is a nice guy who takes care of his bees. Native nectar- and pollen-producing plants like mesquite, cat-claw, huajilla, and Palo Verde once filled these dense and ecologically complex areas where, now, cattle graze. The diverse, natural ecosystem has been pushed out and replaced with grasslands. This new pasture monoculture only produces nectar a few days per year when one legume reaches bloom. What was once a rich, vibrant ecosystem for bees, birds and wildlife has become a weak, dispirited wasteland grazing habitat with limited bee forage. Once, the region was filled with

Native Texas brush country features mesquite (L), compositae, cat-claw, huajilla (UL), and Palo Verde (UR).

undisturbed and diverse forage and bees did well—the survival rate of managed colonies was high. Now, the area cannot support bees year round. What happened? Billionaire ranchers make more money by producing beef for fast-food burgers.

Land use patterns and other factors—including climate change and the shift in bloom timing—contribute to limited bee forage or the acreage of flowers that bees need to sustain a colony. Limited bee forage and extensive monoculture leads to poor colony growth which is often linked to starvation, one of the main causes of colony losses. Other primary causes of colony loss include pesticide exposure, queen failure, and parasitism by varroa mites.

Prior to varroa mites entering the United States in 1987, it was common for beekeepers like my friend Jimmy to expect annual losses of 5-15% of their total colonies. Since then, the national survey indicates that beekeepers lose between 28.8% of their colonies for the past 12 years.

Experts feel that beekeepers should strive for annual colony losses of no more than 15-20%. This loss can be due to natural events that occur in the hive, such as queen replacement failure or extremes in the ever-changing weather. Sometimes, a beekeeper runs into bad luck with a poor nectar flow, varroa mites, small hive beetles, and American foulbrood. While not caused by beekeepers themselves, they must know how to effectively manage each situation as it arises. To keep colony losses less than 15% a beekeeper must receive extensive and accurate training. In the beekeeping world, this means having both stronger, healthier colonies and, theoretically, the financial stability to maintain them.

An Income From Bees Keeps Beekeepers Alive

Making a profit from beekeeping isn't about 'robbing' all the honey from a hive. Experienced beekeepers routinely harvest, or rob, all the honey from the bees. Every year. No exceptions. While they may leave honey in the brood nest, extreme honey removal may not be in the best interest of the bees or the beekeeper.

Other beekeepers are much more selective. Beekeepers have to be more realistic, and work in partnership with the bees. This

After I opened a drone cell, a female varroa mite ran out. A worker bee took notice.

Multiple varroa mites on drone pupae where longer pupation makes more mites.

means the beekeeper may need to delay harvest to ensure colonies have adequate food reserves to survive the next challenge—like a period of drought, a period of heavy rain, or the ever-recurring challenge of winter. But the beekeeper must continue to make money. Keeping colonies alive means keeping beekeepers in business—not losing more money than they make.

Where bee forage is limited beekeepers must provide nutrition for their bees. To replace honey and pollen they must purchase carbohydrate and protein. I know larger commercial beekeepers who spent over $100K monthly to make food patties just to keep their bees ready for pollination contracts. They have moved to a southern location, investing in real estate, in flat-bed trucks, diesel fuel, dozens of employees, and do whatever else is required to move the bees to regions of the country where forage is abundant.

Then they are either ready for pollination or honey production. It's a complicated arrangement, and most beekeepers are constantly trying to improve their management scheme. Remember, all these expenses significantly reduce or eliminate profit to the beekeeper. Beekeepers must not lose more money than they make, walking the line between abundant food reserves in the hives and starvation; between a good income from honey production and substandard colonies for pollination.

All beekeepers, even those with one or two colonies, are advised to view their beekeeping adventure as a business in order to profit from their investment. The inspiration of seeking a profit from bee colonies should encourage, motivate, and promote good beekeeping practices, which means knowing the biology of the hive, not simply taking honey from the hive without knowing what is going on.

Commercial beekeepers are the primary influence on the beekeeping industry, currently driven by the enormous influence of almond pollination. In 2019 the California almond industry reported over 1.1 million acres in production, requiring nearly 2 million honey bee colonies, three-quarters of the colonies in the entire country! The average colony each rented for just under $200, contributing nearly $400,000,000 to the beekeeping industry. In 2020, an additional 150,000 colonies will be needed for new trees already planted that will be entering production.[5]

Meanwhile, the beekeeping industry has increased its diversity of products. The most common source of income for beekeepers is from honey sales if there is a surplus of honey produced. Even a beekeeper with just a few colonies can sell surplus honey to support their beekeeping. Other sources of income can come from selling beeswax, propolis, pollen, or bees and equipment in the form of small bee colonies, or nucleus colonies.

New honey. The white comb has an air space under the wax capping. 'Wet' areas appear where honey and wax are touching.

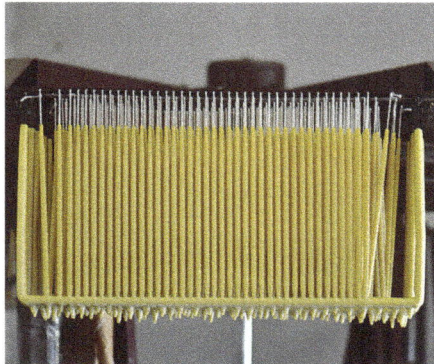

Dipped candles made of pure beeswax being made at a monastery. Once cooled, the candles are trimmed to a specific length and shipped world-wide.

You can be creative. A friend of mine sells a slurry of honey, propolis extract, and pollen called ProPolMel. While making no health claims, he markets this concoction for improved personal health. Others have worked with food manufacturers to develop a private brand of honey-grilling sauce, ketchup, honey-sweetened soft drinks, or energy bars. The potential is huge.

Whatever the means, a newbee may expect to generate income from their new colonies within two years of starting bee management. It is not easy, and it is certainly not guaranteed. Still, it is worth considering. If second-year beekeepers sell 50 pounds of surplus honey (per colony) at $10 per pound, they will generate a gross income of $500 per colony[6], generating enough cash to pay for bottles, labels, medications and replacement queens and make a profit. If a beekeeper successfully makes and sells a nucleus colony, the potential income range is $150-$250 per colony.

Propolis coats the entrance of this nest in south Texas. Comb can be seen inside. Bees remove loose bark and use the resin and beeswax mixture to protect legs and wings.

Tyler Andre and Charlotte Hubbard put honey comb built in a hollow wall into Langstroth frames.[7] C. Hubbard.

But don't expect income to sustain your colony during the first year. Let your bee colonies grow and prosper without harvesting honey, bees, brood, or anything else. Check that the bees are healthy and thriving with abundant food reserves before taking anything from them. One way is to leave the honey on the hive in the fall of the first year and wait till the next spring to harvest the remaining crop. This delay serves as insurance to keep the bees well fed over winter and helps the new beekeeper understand how much food is needed for colony wintering.

In order for a beekeeper to keep both bees and the beekeeper alive, they need a plan. We will explore simple, moderately conservative bee management techniques to support the bee colonies and, in the long run, return some of the investment we put into our colonies.

Developing a Beekeeping Plan

Having a plan[8] is crucial to anything involved with beekeeping, from figuring out where to set up the colony to where to sell your honey. The most important thing to keep in mind is what works best to help the bees.

Before you do anything, there is much to learn. Find a mentor. Join a vibrant bee club. Read a variety of highly rated bee books. Watch videos from industry experts. Listen to the best pod casts. You will need to work on your beekeeping education and study honey bee biology. This will take time—a year or two is advised—with investment in training programs, bee schools, the right sort of beekeeping books, YouTube videos, and some shadowing and mentoring with experienced beekeepers.

Mentors can be hard to find in some areas. If you can, attend all the bee schools where you can volunteer. You may find a good mentor at these events, and then set up a date and time for you to visit them in their operation. Don't expect the mentor to travel to you unless an agreement is arranged, and they receive, at minimum, a good meal or a nice bottle of scotch.

The purpose of finding a good, accessible mentor is to learn beekeeping hands-on, under local conditions. Things that will

Three boxes of Langstroth equipment, eight-deep frame version. This is what I use, favoring the smaller size for bee behavior and development. The frames are interchangable with the five-deep frame nuclei colonies I also use as support hives.

work on the other side of the country may not work for you. A local mentor can talk about local nectar and pollen flows and prepare you for climate variations. This can help you decide, among other things, what kind of hive will work best for the bees in your region.

As you make your journey into beekeeping, information slips through the crack in the first few years, or maybe you are unable to fully experience all situations. For example, the first few years are meant to learn about what is going on in the box with the bees, then with pests of the bees, and then, learn more about the interactions with the environment.

Start your beekeeping career using Langstroth-style bee equipment, not just because it is the industry standard, but most books and information that are readily available discuss Langstroth hives. Once you learn the basics and know what works best for the bees, you might be inclined to try other hive types like the Kenyan Top-Bar Hive or Warré Hive. I recommend holding off on these alternative systems until after you've used the Langstroth for at least three years.

The exception would be where your find yourself in an area where alternative hive types are widely used. You may have better luck finding a knowledgeable instructor or mentor who will work with you to provide thorough training.

Avoid highly promoted speciality bee hives that make it seem too easy to be true. Two that are misleading are the wood and plastic Flow hive and the single box hive that has you place empty jars on top. While they work well for some beekeepers, they may result in colony death in the hands of an inexperienced beekeeper.

Grafted queen cells showing abundant royal jelly in the base of the cup. Localized queen production is a potential source of income for beekeepers.

Delay speciality equipment until you better understand basic management concepts.

Consider starting with four hives. Previously, I had recommended two hives[9], but starting beekeeping with four colonies rapidly expands your level of beekeeping experience, allowing you to quickly see how colonies develop differently under the same conditions. It will give you a sense of responsibility, stimulating you to prepare more by attending courses, reading more books, and finding a good, reliable mentor. Having four colonies gives added you resources to equalize hives, replace queens, and make up losses.

I had the opportunity to train with some amazing professionals— professors and beekeepers to whom I dedicate this book—and you will meet a few of them in these pages. There is still more to learn about our favorite insect. It is a challenge for newbees to learn so much information, apply it, and be a successful beekeeper.

If you have some fun and success with bees and beekeeping, I will have done my job. As you master the key aspects of bee biology and the management of bee colonies, you will integrate your training with what you observe in your colonies. You will see how differently individual colonies will behave to the same treatment and beekeepers have highly variable opinions of what has happened. Do not let this frustrate you. Let it excite and stimulate you to figure out what may have been different. This will be key to growing strong bee colonies and keeping them alive.

Let's get to work.

Lawrence John Connor
June 27, 2019
Kalamazoo MI 49001 USA

1. EDUCATION & PRE-TRAINING

Three Methods x Three Ways Approach to Beekeeping

Each person who considers getting into beekeeping has a choice. Jump into a sea of conflicting lore and hope their bees live, or prepare in advance and immerse themselves in facts and proven practices of beekeeping and be ready for whatever the bees give them.

When you start beekeeping, you really have no idea what you do not know. It would be wonderful if every new beekeeper, 'newbees' as some call them, could find success keeping their bee colonies alive. The reality is that probably one third of all new beekeepers give up after two or three years of trying because their bees keep dying.

Too often, beekeepers lack proper education and fail to do the job properly. My solution is to use "three methods three ways". That means that you should use *three styles of learning* and *three versions of each one*.[10]

An excellent training program could involve a minimum of three good beekeeping books, plus three good bee courses, plus three fantastic bee schools. That's a pretty good place to start. Get a wide range of experiences rather than following just one instructor, professor or writer, not only adding to your knowledge, but diversifying your sources.

Reflect on the thriving bee colony where there is a great diversity of genetics because many drone fathers have mated with one queen. The result is a colony with more genes to control both big and little issues. As a beekeeper you need a similar diversity, with more tools. It is not just the act of spending a few hours reading a small book on beekeeping, but putting in hundreds of hours to learn about bees, talking to experts, questioning folklore and finding accurate information sources.

Three Book Sources of Information

Printed, electronic and audio books are commonly used to establish a solid foundation of beekeeping knowledge. As a writer and publisher, I would, of course, love it if you read all my books and those published by my firm, but it is important that you don't read *only* my books. I just said that is not a good idea. It will limit you, and there is much more to learn than I can teach you. You need a variety of sources for your basic information.

Here are three groups of books—choose at least three to start your training platform. They differ in their treatment and intensity, and they use different methods and focus on different areas. They have different levels and types of inspiration for the new beekeeper. Some use facts, science and data as a motivator, while others motivate the reader with the amazing biology of the bee.

Select three titles—one from each group—and you will be exposed to different tones, perspectives, and ideas. There are disagreements in any field, and beekeeping is no exception. This allows you—as a new beekeeper—to make informed decisions based on a wider set of sources, fit to your own situation. Record the topics of agreement and conflict and ask others for their views on these issues.

The key to your reading is to build a solid platform of varying opinions and experiences. Like a three-legged stool used to milk a cow, you should establish a solid base of your information about bees and beekeeping.

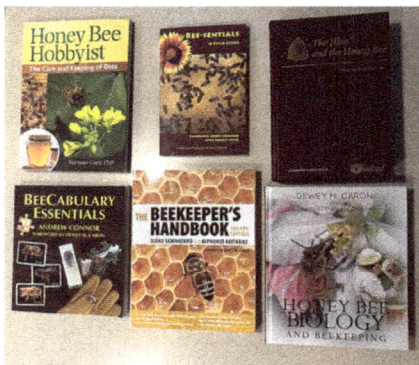

Three groups of books for pre-beekeepers, new beekeepers and more advanced beekeepers. See the list on page 27.

Pre-Beekeeping Introduction and Learning About Bees:

Gary, N.: *The Beekeeping Hobbyist*, Companion House Books
Connor, A.: *BeeCabulary Essentials*, Wicwas Press

First Year Beekeepers

Connor, L.: *Bee-sentials: A Field Guide*, Wicwas Press
Sammataro, D. and A. Avitabile: *The Beekeeper's Handbook*, Cornell University Press

More Advanced/Reference Textbook

Caron, D. and Connor, L.: *Honey Bee Biology and Beekeeping*, Wicwas Press
Graham, J.: *The Hive and the Honey Bee*, Dadant and Sons, Inc.

There are many other books out there. Carefully consider excellent books by K. Delaplane, J. Tew, R. Bonney, K. Flottum, and a pending book by J. Ellis and a revision of the *ABC and XYZ of Beekeeping*. Check the Wicwas Press LLC website for updates on recommended books: www.wicwas.com

Some classic books to consider to round out your library are hard to find and often expensive, especially those written by Dame Eva Crane: *Honey: A Comprehensive Survey (1975)*, *The Archaeology of Beekeeping (1983)*, *Biology and Bees and Beekeeping: Science, Practice, and World Resources (1990)*, and *The World History of Beekeeping* and *Honey Hunting (1999)*.

On the subject of bee biology consider the classic 1987 work by Mark Winston, *Biology of the Honey Bee;* or the more recent *A Closer Look Basic Honey Bee Biology (2017) by* Clarence Collison.

An amazing body of work is found in the science writing of Thomas Seeley: *Honeybee Ecology (A Study of Adaptation in Social Life)*(1985), *The Wisdom of the Hive: The Social Physiology of Honey Bees (1996)*, *Honeybee Democracy (2010)*, *Following the Wild Bees: The Craft and Science of Bee Hunting (2016)* and *The Lives of Bees: The Untold Story of the Honey Bee in the Wild (2019)*.

On pollination: consider John B. Free's *Insect Pollination of Crops (1970)* and S.E. McGregor's *Insect Pollination of Cultivated Crop Plants (1976)*.

Photos of the two magazines, *American Bee Journal* and *Bee Culture*. Every beekeeper needs to subscribe to at least one of these publications. Electronic versions are available as well.

On queens, breeding and genetics: Laidlaw and Page: *Queen Rearing and Bee Breeding* (1997). For queen rearing: *Queen Rearing Essentials* (2015).

For plants for bee forage: P. Lindtner's *Garden Plants for Honey Bees* (2014) and T. Horn-Potter's *Flower Power: Establishing Pollinator Habitat,* (2019).

National Monthly Journals

Bee journal subscriptions are not that expensive, and if you get one good idea or fact out of each monthly issue you have received a good return on your investment. Check the sections on new products and follow a favorite contributor—most invite your questions and ideas. Look at the calendar of upcoming bee meetings and plan ahead.

Both magazines are available in electronic editions.

American Bee Journal www.americanbeejournal.com
Bee Culture Magazine www.beeculture.com

University of Guelph apiary, site of Canadian YouTube series.

Dr. Jamie Ellis at the U of Florida apiary, site of IFAS YouTube series. Ellis photo.

Three Video Sources of Information

Several universities have produced extensive videos available on YouTube on starting with bees, the challenges and opportunities of beekeeping and related details. I direct you to these rather than several other sources, which vary in their level of content[11]. Watch as many as you can, take notes, and formulate questions for classroom discussion and lectures. I recommend these because of the reputation of the presenters and their university backing.

The University of Guelph

Professor Ernesto Guzman and Master Beekeeper Paul Kelley have developed an impressive array of very focused video programs, available on YouTube. Get the list by visiting the website for

Students at a community college beekeeping class learn the basics under supervised instruction.

entomology at the University of Guelph in Ontario. This should be carefully considered by all potential northern beekeepers and checked out by everyone.

University of Florida

A variety of YouTube Videos, featuring Dr. Jamie Ellis from the Entomology Department plus students and staff at the University of Florida in Gainesville are useful for all beekeepers, especially those located in the southeastern United States. Some were recorded about a decade ago, but several are newer. The content is top quality with precise science represented.

Beekeeping in Northern Climate Video Series

A series of videos were produced by the University of Minnesota's Department of Entomology Bee Laboratory and UMN Extension. They feature lead UMN apiculture technician and entomologist, Gary Reuter, and are intended to be instructive and entertaining vignettes on a variety of beekeeping topics. Each video covers a single topic, and you do not have to watch them in any special order. In their entirety, they are a lesson on how to keep bees in colder locations, but I suggest everyone watch them, take notes, and compare methods presented from different regions.

Students at Western Michigan University started a Sustainable Beekeeping Program as a campus project. Here, some of the participants sample some of the first honey their package hives produced.

Other states have basic programs for beginner beekeepers. Check out Michigan State University Beekeeping YouTubes. They were posted within the past year. Also, University of Montana beekeeping courses are well represented for different levels of beekeeping, including their training programs leading to their Master Beekeeper certification. Finally, check out videos from Cornell University, the University of California, Davis and others.

Brushy Mountain Bee Farm Videos

Brushy Mountain Bee Supply Co., now out of business, had a strong connection to video education over the years. The collection was a bit uneven, the techniques varied, and some content may need to be confirmed elsewhere. I cooperated on a number of Webinars which are still found through www.wicwas.com. Also check with YouTube for availability of these programs.

Other Sources

There are many YouTube videos other there, and some may be exactly what you need to watch. Others may not be helpful and a few on advanced levels are probably confusing or even harmful for the new beekeeper. I have done a great deal of damage control over the years because of what newbees have seen on the Internet.

Select and review carefully. Do a little research about the speaker and see if they have the credentials you seek. Again, take notes and compare statements you find with other sources.

Three Courses/Bee Schools/Training Programs

Consider active participation in three beekeeping training programs. Consider three as a minimum—my most successful beekeeper friends attend three or more programs every year, not just to keep on top of what is new in bees and beekeeping, but to learn about the new research that has been done, and get new perspectives from an invited speaker.

Many state and most local beekeeping clubs conduct an annual course or class in beekeeping. Seek a minimum of three of these programs. Consider attending your local bee club's annual beekeeping training school, a community college course in beekeeping, and your state organization educational meeting.

A few groups offer Master Beekeeping classes. These vary in objectives, content, organization, rules and fees. Some are designed for new beekeeper training and may take three to six years to complete. Others are clearly not for the new beekeeper—in fact, some require that a participant have kept bees for a minimum of five years and completed a long list of public service and advanced beekeeping topics. Search for courses for people interested in learning about bees and beekeeping as potentially new beekeepers, newbees and novices. Know beforehand why you want to be a Master Beekeeper and how much time and money you are willing to spend. Is it just for bragging rights? Or do you really want to contribute to your community of beekeepers?

Regional Organizations

These three groups vary in their organization, but over the years have all done a remarkable job of bringing together new and small-scale beekeepers for multi-day educational experiences. Review the program and see if it fits into your learning needs or use as a brush up.

Visit their websites and see what they are offering this year— all three groups meet in a different state or province every year, and while the structure often appears the same, the presenters and content changes dramatically every year. I recommend the programs that feature university extension specialists and researchers who are also teachers.

Eastern Apicultural Society www.easternapiculture.org
Heartland Apicultural Society www.heartlandbees.org
Western Apicultural Society www.westernapiculturalsociety.org

National Beekeeping Organizations

Not generally suitable for new beekeepers, but if there is a convention near you, look at the program and pick a day with general and basic talks to attend. You may not appreciate the

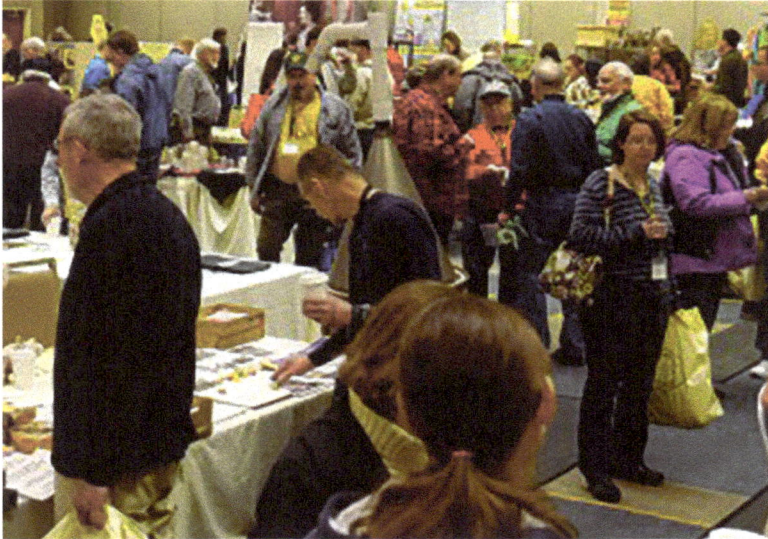

Heartland Apicultural Society meeting vendor area. A good meeting combines stimulating and competent speakers, informative break-out sessions, field sessions (as weather allows) and a vibrant beekeeping supply and trade show. HAS Website.

lobbying, honey promotion and political parts of these conventions. The two national beekeeping groups offer updates on bee research, national beekeeping politics, and social programs. Their offerings are usually less useful for the brand-new beekeeper.

American Beekeeping Federation www.abfnet.org
Amer. Honey Producers Assoc. americanhoneyproducers.org

Beekeepers in Canada will find a wide range of local bee club sponsored meetings. Also consider the Introduction of Beekeeping course offered by the University of Guelph, outside of Toronto every spring. Dr. Guzman and Paul Kelly run the very popular Introductory Beekeeping course for both beginner and established beekeepers seeking more in-depth knowledge about bees and the beekeeping industry. The course is offered the last weekend in April or the first weekend in May. The limited enrollment course is held onsite at the Honey Bee Research Centre, Townsend House at the University of Guelph.

Various programs may be offered through provincial apiary and beekeeping associations. I have participated in top-quality programs in both Alberta and British Vancouver. Dr. Medhat Nasr, retired Provincial apiarist, offers training in Alberta.

Above: A member of the San Francisco Bee Club keeps personal colonies at community gardens, helps train new beekeepers, and donates some of the honey to city soup kitchens.

R: Dr. C.C. Miller gave up his medical practice, which he did not like, and kept bees for many years.

Mentoring/Shadowing

I again advise every person considering beekeeping take the effort to locate a qualified individual who is able to serve as a teacher, mentor and beekeeping coach. Offer to do some routine work, like nailing frames or extracting honey, in exchange for training. Or offer to pay a fee to be a student of the experienced beekeeper. Try to do as much as the beekeeper does, carefully following his or her supervision.

With shadowing the participant rides along and watches what the beekeeper does and discusses what and how things were done. The newbee may not perform any bee work.

Once a season of beekeeping working with a mentor has been completed, you should have a plan for getting your own equipment and bees. Keep the habit of attending classes and working with knowledgeable individuals for your lifetime in beekeeping.

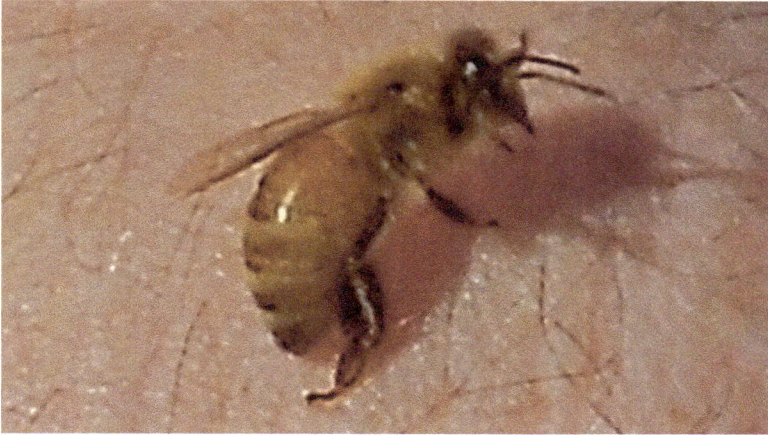

Bee stings are part of beekeeping. So are hot, sweaty days performing overdue chores in the rain and feeding colonies on a cold winter day. These days are balanced by seeing your queen for the first time, removing your first full super of honey, and tasting honey that you and your bees help produced. Together. D. Caron.

Options To Keeping Colonies

After you receive this extensive training, you may decide that beekeeping is not what you want to do. Maybe it is a matter of timing, living situation, physical limitations or unexpected finances. But it is okay to say that you are just not that into the topic, and you feel you should not start a hive of bees.

Some people are afraid of being stung—some with good reason if they have a history of an allergic reaction specific to honey bee venom. Others realize that their current living situation is not appropriate for bees. Hives filled with honey are heavy, and honey harvesting makes a mess. Consult with your family or support network to determine their level of support, as well as their willingness to help produce and sell your bee's products. I am recommending all new beekeepers start with and all current beekeepers maintain a minimum of four colonies as a means of keeping resources available within their apiaries and potentially increasing colony survival rates. This is a big expense, and if you cannot afford it, you should reconsider keeping bees.

You do not need to keep bees to help them. Study native bees and other pollinator species. Teach folks about bee-friendly gardening, and what trees and shrubs to plant to help feed bees. Join the effort to plant wastelands and toxic sites with nectar and pollen producing trees and shrubs. Help at local bee schools with registration or host a speaker. Get into the classroom and share the world of bees with students and teachers. Most of all, have some fun.

If you are afraid of facing colony losses, keep in mind that many very successful beekeepers have had major colony losses. C.C. Miller, MD., wrote in his book *Fifty Years Among the Bees* (1915) about his losses during the 1871-1872 winter: "*By April I had only three colonies left, two of which I united, making a total of two left from the forty-five or fifty.*" This did not force Miller to give up, as he continued to buy more bees and increased his colony holdings. This is not an uncommon story when men and women find joy in beekeeping.

2. APIARY SITES

Evaluate Your Potential Apiary Location

Once I knew a clever beekeeper who placed 20 bee colonies on a used trailer and moved them into cucumber fields for pollination. The pickle growers, as they are called, paid him just enough to make this work. They were efficient growers, using large machines to harvest the pickles. Knowing how many acres they could harvest a day allowed them to stagger plantings. Because they wanted all the cucumbers to be the same size, they planted so all the flowers in a certain field bloomed at the same time, and a few days later all the cucumbers were ready to harvest. The staggered plantings bloomed throughout the summer, setting up a steady flow of same-sized harvested pickles into the processing plant. When the cucumbers were ready for harvest, a large machine pulled the plants out of the ground and sent them through a series of rollers that popped the cucumbers off the vines. Because the

View of overwintered 8-frame colonies in SW Michigan. The barn provided both wind protection and equipment storage. Bees and Christmas trees were a big part of my life on the family farm as a teenager and bees again in the 2010s.

bees needed to follow the blooming flowers in different fields, the mobility of the trailer saved the beekeeper considerable hand lifting of colonies.

At the end of the pollination season, the beekeeper took his trailer back to his permanent apiary. He settled his colonies on the ground and prepared them for winter. He fed them (because there is little nectar in cucumbers), medicated them if necessary, protected them for winter and monitored them carefully until he would go back to the pickle fields the following season.

While the initial investment for the trailers was expensive, the setup lasted this beekeeper a number of years. When all was said and done, the beekeeper did what he had to do to keep both the bees and his pollination business alive.

Today the granddaughter or grandson of this beekeeper probably uses a fork lift and a two-ton truck to move bees in and out of the fields. The scale of these capital expenses make commercial beekeeping an expensive business.[12]

The mobility of the hives also allowed for a rapid removal of the colonies in the event of unpredictable hazards like a fire, floods, and even hurricanes. The most common reason to move bees in

Unloading package bees at Heitkam's Honey Bees in California before shipping to Kansas. Using fork lifts is a routine part of commercial beekeeping.

Florida three-frame mating nuclei provide new colonies and queens.

a hurry is because of an insecticide spray applied by the grower paying for the pollination.

I don't recommend new beekeepers consider trailers for a mobile apiary. In fact, very few should even consider that as an option unless planning to use colonies for pollination. The guidance of a mentor may help you carefully evaluate where they plan to place their colonies. As you develop your plan to obtain bees and start colony management, you should carefully evaluate all your potential apiary locations, keeping a few important things in mind including potential hazards and benefits.

As you start out, invite a mentor or another experienced beekeeper to help you evaluate your options before you place bees in a new location. They may help you avoid a potential disaster, like putting bees in a location where horses run free, on the edge of a lake or brook that floods once every few years, or next to a school playground.

Use every tool you have to look at the area around the apiary. Access to maps, including satellite images,

Ted Miksa works with his mother, Linda Miksa, on cell builders together, sharing conversation and, if available, an extra frame of brood or pollen.

is helpful to see what is within a two, three and even four mile radius. In Box 1 you can see how much more foraging acreage is available to colonies as foragers fly increasing distances.

Every year reevaluate your apiary location. Has honey yield increased or decreased? Are you successful in making increase nucleus colonies and overwintering them?[13]

There are plenty of other factors to consider, but first, let's start with a consider different weather patterns.

Seasonal Considerations

Certain locations may be very good for summer nectar collection but may not be good for wintering. You will probably need to visit the colonies in the winter to evaluate their hive weight for remaining honey and feed as necessary. If you have deep snow, use snowshoes and a sled filled with winter sugar patties to put on the bees to prevent starvation. The ideal location should be good year-round, and not require hive movement.[14]

It's not uncommon to find beekeepers who have been stuck in snow, mud, *and* blow sand travelling into and out of the same apiary over the course of a year. If you move bees to a different part of the country, that nice flat spot you found in a open area may become a raging wash as snow melt or rain comes off the mountains miles from away. Top heavy colonies blow over in

Having a younger person able to snowshoe into a winter location, pulling a sled filled with feed and supplies, is a lesson in location selection and mentorship. T. Jones.

70 mph winds and are torn to shreds by flooding rivers. Prior planning is essential.

Wind

You want to avoid locations with extreme wind exposure. Colonies placed in direct exposure to the line of prevailing winds consume more honey to survive. Find a protective location out of the wind. Visit the area on a cold but sunny day, and take off your jacket. If you're comfortable, the bees will probably be comfortable as well. If not, it may not be good for the bees, so rethink putting bees in that location.

Strong storms blow lids off colonies and knock over teetering, top-heavy hives. Make sure the colony is solid on the hive stand, and will not go rocking in the wind and fall over. Straps and fence posts are standard preventative tools in these conditions. Use a big rock on the lid to keep it from flying.

New nuclei colonies with bricks to keep the lids from going airborne.

That shady site becomes a mess if trees fall during a wind storm. T/B Jones.

Sun and Hive Temperature

Many beekeepers place their colonies in locations where the bees receive full sun all day long. Full sun keeps colonies warm and dry and discourage two pests: *Aethina tumida* (the small hive beetle) and *Varroa destructor* (varroa mite). These pests are sensitive to low humidity and stop reproduction when in hot and dry conditions. Remember, honey bees evolved in the tropics and can handle heat and dry conditions better than either of these pests.

Open and sunny locations provide easy access and lower mite and beetle numbers. D. Caron.

Early morning sun exposure warms the hive and stimulates the bees to work earlier in the day, theoretically leading to greater productivity. The entrances of natural colonies found in bee trees in the wild often face south and southeast. Bee colonies seem to do best when their entrance is warmed by the morning and midday sun. If you cannot arrange your hives this way sun exposure may not be a reason to reject a location.

Fire

Fire is another concern, especially in western states and regions with routine drought. During extremes in dry weather, there is always the risk of a forest or grass fire. Keep bees out of extremely high-risk areas and put bees in less risky locations such as in the middle of a fire lane or vegetation-free region to reduce the risk of fire reaching the hives. Even suburban, small-scale beekeepers lose colonies when a planned burn gets out of hand and destroys their colonies. New plantings of prairie plants require periodic burning—keep this in mind when putting bees in such a location.

Generally, keep colonies away from vegetation by placing them in an area of bare ground, and avoid any location where trees may flame up and fall onto the hives.

This grass fire came up to the hive before it was put out. The bees were busy fanning to cool the hive.

Ted Jones moves colonies away from a flooding stream. B. Jones.

Flood Water

Many potential locations run the risk of high water from flooding following snow melt and heavy rain. During dry periods, there may not appear to be much threat of high-water danger, but in areas where droughts have been routine for a number of years (during which beekeepers have set up and operated bees for several seasons), a return to 'average' moisture level will increase the chance of flooding from a local stream or lake. Moderate flooding where the bottom board is covered with water can be tolerated if the bees have a good upper entrance, locations where water rushes through and carries hives down the river must be avoided at all costs.

Bears and Bear Fences

Bears are especially fond of bee brood and honey and will actively seek out hives in their home range. Hives should be located as far as possible from timber, brush and cover to exposing bears on their travel routes.

This bear came up to the fenced yard as the beekeeper worked hives. T. Jones.

Jerry Repasky inspects bear damage to his western Pennsylvania apiary. S. Repasky.

Electric fencing has been shown to be nearly 100% effective in deterring bear damage when set up and managed properly unless the bear has success. Then the creature will return frequently, fence or no fence. Compact apiaries are easier to protect with bear-resistant fencing rather than those scattered over a larger area, so beekeepers should consolidate hives to form a single apiary that can be practically managed.

If you do suffer bear-related beehive damage, check with your local cooperative extension office—you may be eligible for some monetary compensation to replace your damaged equipment and bees. Also, in some areas, financial assistance is available for setting up fences to deter bears.

Vandals and Hive Thieves

There is no explaining human behavior when discussing the out-and-out destruction of property. Like cow-tipping and bashing mailboxes with a bat, certain individuals find bee hives a challenge to destroy. Driving over hives with trucks, tractors, snow machines or poisoning hives with insecticides force beekeepers to find isolated locations where vandals cannot find the bees, or put the bees directly right in front of property owner's line of sight so they will see any activity going on in the apiary.

Theft of hives has become very common as the value of pollination colonies has increased. This is an annual problem in the almond pollination areas of California. One approach is to embed electronic sensors in the hive bodies and pallets, making the hives easier to identify and even locate. Traditionally all boxes and frames were branded—the wood was scored with a hot iron made into a unique logo or identification mark for the beekeeper. Many states require that hive bodies be branded or labelled for quick identification of ownership.

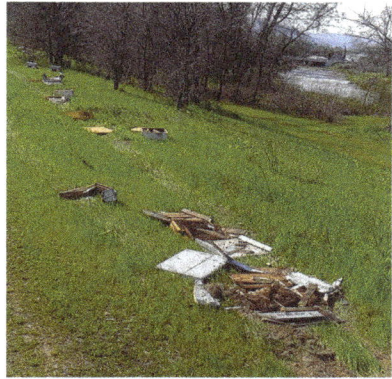
A man driving a black truck was observed driving over these hives. Internet.

Pesticide Exposure

As we learn more and more about what causes deaths in bee colonies, we learn that there are many compounds being used in the environment that negatively impact the colony, outright or indirectly. Even the water coming off intensely treated lawns and golf courses may be toxic to bees obtaining a drink. When setting up your apiary, check for routine use of chemicals in agricultural crops, around golf courses, on the lawns of corporate headquarters or anywhere pesticides are used.

Researchers are becoming increasingly aware of how certain pesticides cause sub-lethal damage to the bees when they used together with other chemicals. These are called synergistic effects. Certain fungicides impact bee brood and developmental rates, while insect growth regulators are known to affect colony health.

Students pinned and mounted the pollinators killed when two *Tilia* trees were sprayed in full bloom. Internet.

45

Dead bees of a strong colony apparently killed by Seven insecticides. R. Burns.

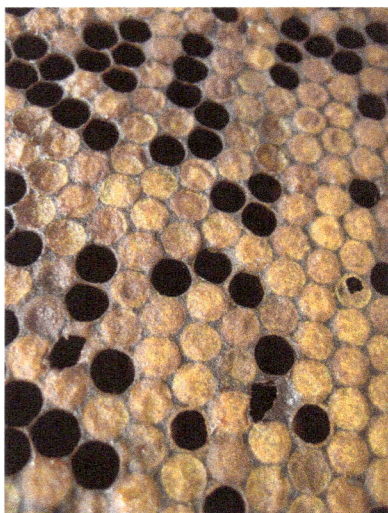

Sunken cappings of cells containing pupae killed by American foulbrood. Look for darkened cells and uneven holes in the sunken cappings. AFB has a strong odor to many people.

Disease Exposure

Speak with your local or state/provincial apiary inspector to see if your potential apiary location is near an area with a history of chronic American foulbrood infected colonies. This disease is potentially very harmful to bees because the bacilli form spores after they destroy a larvae. These spores last for decades and easily infest colonies installed into the area. The spore source may be from dead bee hives—managed and feral—and serve as a reservoir. Avoid such an area if you wish to have a sustainable beekeeping operation.

Hazards to Humans and Animals

In selecting an apiary location, carefully consider others who will be working and playing in the area, especially children, the elderly, those with limited mobility, and animals in confinement. No apiary should be established where there is a probability of doing harm to humans, animals, or the bees themselves. Some humans are

These bees are returning to their hive following honey harvesting.

easily intimidated by certain bee behaviors, making it necessary to keep bees away from as many people as possible.

For example, when bees are young, they make cleansing and orientation flights around the front of the hive entrance. There are often a large number of young bees that do this at the same time. Also called 'play flight', all bees, including young queens and drones, make these flights in order to empty their rectum and learn the location of their hive. For some people, the sight of several hundred bees flying in front of a hive can be unsettling. Then they discover the deposits of fecal material on their car, camper or clothing on the line and become upset, reacting unpredictably toward the bees or the beekeeper. Or both. Also, inexperienced beekeepers who have not seen this flight behavior may mistake it for swarming, which it is not. When non-beekeepers see this behavior, they might think that an angry bee attack is about to

Swarm passing overhead in apiary at the farm. They may seem intimidating but are quite harmless.

47

start. These bees are harmless. These bees are no more likely to sting than bees foraging on flowers.

Bee flights are part of foraging and mating behavior, but it may interfere with human activity. During normal food gathering, worker bees fly out of the hive and head as directly as possible to the food source, which has been communicated to them via their dance language. This may take the bees right over a sidewalk, your neighbor's lawn or garden or through a pet area. An open suburban lot filled with foragers going in and out of the colony may disturb a neighbor. A simple fence, fast-growing hedge or even a barrier made of burlap strung between fence posts directs bee flight up and out of the way.

While you may think it looks pretty to have your hives at the edge of your flower or vegetable garden, heavily loaded bees returning to the hive may fly over the garden where you are working and may get tangled in your hair, causing an accidental sting. This is not the same as the bees foraging on your garden cucumber flowers; those bees rarely cause problems, and you may not even notice they are there.

What to Look For in an Apiary Location

Water Source

As temperatures increase, the overall percentage of bees collecting water increases. Always make sure to supply bees with water, either from a dripping garden hose, a water feature filled with tropical plants or a jar filled with water in the hive entrance (Boardman feeder). A clean water source should be within half a mile of hives and preferably within the apiary. At temperatures above 105°F, most of the worker bees in a colony are primarily collecting water.

A water feature filled with tropical plants establishes an attractive water source that you can enjoy as well, especially in your back yard. Since beekeepers have discovered that bees prefer water with minerals and salts, a few mixed rocks and mineral stones in the watering device may benefit the bees.

Water feature provides hundreds of bees water to carry back to their hives.

Nutrition

Additionally, factors like nutrition or survivability hazards, including those caused by humans, wildlife, and nature, will affect your apiary site. Your primary motivation in placing colonies should depend on the amount of available bee forage throughout the season. Determine whether or not a particular area is adequate in terms of food production through the floral distribution and geographical features of the area.

Wherever you have an apiary location—in the country or on a roof in a huge city—some locations may be very good for spring buildup and early nectar and pollen, but then turn into a food dearth that force the bees to consume their stored food as well as steal food from each other. This may suggest that you need to move the bees for a better location for the rest of the season or adapt a feeding program to sustain the bees when natural food is not available.

Mustard flowers are a rich source of pollen and nectar.

49

Box 1. Acreage at Different Distances

Radius (miles)	Number of Acres
One Section[15]	640
1 mile	2,009
2 miles	8,049
3 miles	18,111
One Township	23,040
4 miles	32,198

At 4 miles, there is remarkable evidence of the amazing and highly refined scout bee foraging behavior of honey bees, in that the scout found one field, perhaps just 100 acres out of 32,198, returned to hive to recruit foragers, and the bees visited the field in numbers benefiting the hive. 100/32,198 is just 0.0031 percent of the real estate available within this 4 mile radius.

Map out your proposed apiary location and draw circles at one-mile intervals. See how many acres the bees will cover at these increasing distances. Beekeepers must realize if just one neighbor mows the dandelions not much forage is lost. But if all homeowners within 4 miles do the same thing, the loss of food reserves is enormous.

Move the center of your radius—your potential apiary site—by as little as a quarter mile and observe how much your foraging range changes. This may put you into a richer nectar producing region, or it may expose you to heavier agricultural pesticide usage.

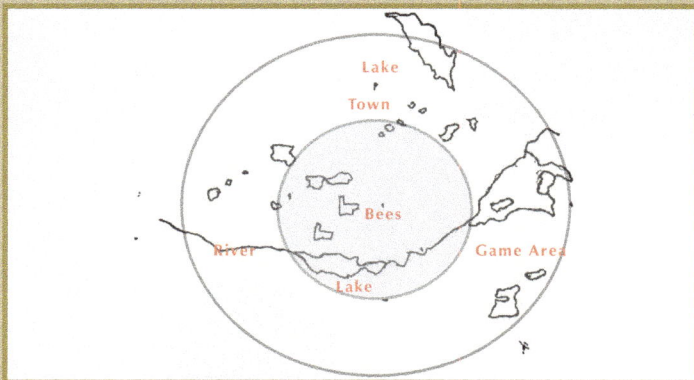

The farm was located with forests and small farms. Several lakes and a river were located in the 2 mile radius (inner circle). The outer circle is a four-mile radius.

Returning pollen foragers showing the different plant sources they have visited.

The blue pollen color is from early spring blooming *Scilla*. Bees kick the pollen into cells with their hind legs.

Understanding Bee Foraging

The biology of bee foraging reflects the abundance and diversity of food a colony seeks to fill its nutritional needs. Rarely does all of the food come from one pollen or nectar source at a particular time. One colony may have foragers collecting food from 20 or 30 different plant flowers. A second colony, perhaps situated right beside the first, may be gathering food from a different group of locations and plant species.

These foraging patterns change daily, if not by hour, as flowers come into bloom, reach peak pollen and nectar production, and then fade, leaving only a few late bloomers to keep the bees occupied.

Not surprisingly bees learn time based on the blooming cycle of the flowers they pollinate. Flowers that open at unusual times of the day or evening are excellent examples. This is dramatically shown in the evening blooming cacti and *Datura*[16], where flowers are visited by bees in a very narrow window of twilight, just before full darkness, when new flowers have opened, but before full darkness is reached. Otherwise the flowers are pollinated by night flying insects like the moths and animals, such as bats. While these flowers may remain open into morning, they often close quickly as their pollination needs have been met. And if all the floral resources, pollen and nectar, have been removed, morning pollinators are likely to find little to collect.

Honey bee scout bees usually include some of the oldest and most experienced bees in the colony. They constantly search for better food sources, even during a major nectar flow. They may find usable resources near the hive location, or they may find something really rich in sugar or protein several miles from the hive. When she returns to the hive the scout bee does a bee dance (round or wag-tail) and recruit foragers to collect the food. The foragers evaluate the food samples presented by the foragers, decide which is better and go to the most productive site.

Bees forage in all directions for miles from the hive. This fact is often lost on new and even long-term beekeepers, who look out over a newly mowed lawn of dandelion and moan the loss of food reserves for their bees. While it is a loss, bees are always searching a much wider and greater distance from the hive to find food. In Box 1 we see the role of distance from the hive (radius) and the increasing number of acres enclosed by a growing circle.

Nutrition from Flowers

It confuses new beekeepers to see vast expanses of flowers in full bloom and not a single honey bee visiting them. Honey bees do not visit every species of flower that nature produces. For example, honey bees rarely visit flowers in the tomato family because the flowers produce little nectar and the bees lack the ability to buzz-pollinate the flower, a technique the bumble bees use to obtain pollen. Sometimes, the flowers are covered with pollinating fly species and, other times, there may be a dominant species of

Honey bee collecting pollen and nectar on dandelion. S. Way.

Pollen stored in comb next to the brood. File

native bee on the flowers. This is where a good teacher/mentor is so useful to the new beekeeper.

A mix of nutritional pollens is needed for the production of bees. When well-fed as larvae, the bees produced going into winter are called winter or fat bees—their cells are filled with stored fat and protein that helps them survive a winter or dearth period. These bees facilitate the production of nutrients for brood rearing.

For the sustainable beekeeper looking to master their beekeeping skills, understanding plant species—even varieties or cultivars of certain plants—is extremely valuable for greater honey production or pollination. Since the honey bee is not native to the Americas, it must compete with native pollinators but usually coexist with them without causing harm.

Many agricultural and garden flowers have also been introduced to the Americas. Some, like apples, apricots and oranges are not native to North America but have been widely embraced by human culture. Other species, like spotted knapweed (*Centaurea* spp.) and purple loosestrife (*Lythrum salicaria*), are highly regulated in some areas and subjected to biological control agents. Since honey bees are generalists—visiting a wide range of species—they pollinate

Many excellent pollen and nectar sources are considered invasive species and under removal efforts. This is purple loosestrife. Note the dark blue pollen.

this floral smorgasbord as efficient pollinators while benefiting themselves by filling their hives with surplus honey and pollen.

Beekeepers should become part of the conversation about the policies surrounding the removal of so-called invasive species. In areas where honey bees, and other bee species, are benefiting from certain plants, this benefit must be balanced against the concerns of ecologists, farmers and ranchers who have their own agenda at work.

We know that not all pollen sources have the same value. Learn which pollen and nectar producing plants are important to bees and general colony management plans for your area. A starting reference for the relative value of some honey plants is Peter Lindtner's *Garden Plants for Honey Bees*. The author ranks each species by its pollen and nectar attractiveness to honey bees.

How Many Colonies in One Location?

Your community may set a limit to the number of colonies allowed on the property, especially in urban and suburban areas. In rural areas, there are generally no restrictions.

No matter what level of experience you have with beekeeping, the number of colonies in your apiary should be determined by the food potential of the site. I recommend that all new beekeepers set up a minimum of four colonies their first season but not more than 20-24 colonies in order to minimize potential losses, as this is a huge financial risk for a novice beekeeper to take.

Garden plants like this Indian blanket are excellent food for many pollinator species.

A single honey bee taking a rest. Perhaps a scout bee, the oldest of the forager bees. A. Connor.

If you start with a smaller number of colonies as you develop a new yard location and build your colony holdings over time as experience shows that the site can support more bee colonies.

An overgrown area in the corner of your property may be the ideal place for hives. There are colonies behind the bushes.

Another way to increase your colony's survivability level is to lower your colony numbers if the area cannot support your colony numbers. The late Brother Adam, who bred the Buckfast Bee in England, used a rolling average for honey production. Because of the wide range of variation in honey production for location in southeastern England, he used the average honey production per hive for the previous 10 years.[17] As a new year was added, the oldest figures were removed from his computation. This should generate a trend chart, which Adam advocated should be moving upward as a result of improved stock.

Learn to adjust colony numbers to maximize efficiency for the nectar flow based on production data in a particular area. There are sweet clover locations in the Dakotas that can handle 80 or more colonies quite well but only during the yellow sweet clover bloom. Yet in a heavily forested region of New England, I have seen apiaries with as few as four hives. In other locations, I find that bee yards containing between 12 to 30 colonies are routinely possible. Yet, if there is a great deal of nectar and pollen present, apiaries of 40 to 60 or more colonies are possible during the limited time of nectar production.

Very large apiaries of 100 or more colonies are suitable for major concentrations of dominant nectar producing plants; these colonies are moved into and out of the area for the flow and not left in one location all season. This is typical of commercial beekeeping locations.

Start each apiary with the number of colonies you think it can support. As you work your bees, adjust this number as you see how productive the location is, and evaluate how well you are able to handle the workload.

Pollen is a critical part of colony health and survival. A 2019 study sampled pollen production in nearly 394 urban and suburban apiaries in four regions of the country: Texas, California, Michigan and Florida.[18] See what is produced in areas close to you.

Posted yard in northern Texas.

Commercial mating apiary on the big island of Hawaii, where neighboring hives managed by other beekeepers number in the thousands.

3. MITE TOLERANT STOCK

Necessity of Using Mite-Tolerant Bee Stocks

Perhaps the most compelling component in our objective to keep bee colonies alive is maintenance of mite-tolerant queens. Selection of mite-tolerant bee strains benefit the beekeeping industry by reducing the amount of chemical treatment needed to control varroa mites, one of the most important challenges to keeping bee colonies alive today. By using tolerant stock[17], I know I have reduced contamination of my hives.

Over the past decade I have worked with several different mite-tolerant queen strains. Some were naturally mated, and several precious ladies were the product of instrumental insemination. They survived above the 80% level and were productive. Their colonies did not require miticide treatments. After a decade of working with these queens and their daughters I am convinced the use of mite-tolerant stock is the key component all beekeepers must incorporate into their apiaries to keep colonies sustainable. Of course, proper management must accompany this practice.

Natural Tolerance[19]

The varroa mite, *Varroa destructor,* evolved on the Eastern honey bee, *Apis cerana.* The Asian species and the ectoparasite evolved a relationship where the mites reproduce only on developing drones. When the Eastern and Western honey bee species were kept in the same parts of the Asia, the varroa mites infected the Western bee and were extremely destructive. The mites reproduced on both drones and workers, killing huge numbers of managed colonies and many feral colonies. Fortunately, a few Western honey bee colonies survived and evolved tolerance mechanisms to the mites. The mites had moved onto a new host.

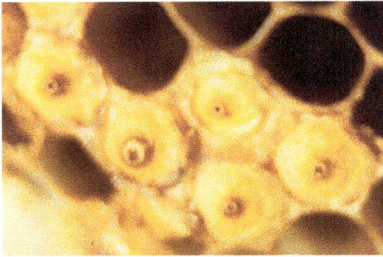

Apis cerana drone brood. where varroa mites evolved. G. Koeniger.

Populations of bees have developed several mite tolerance mechanisms, without human help. These include *hygienic behavior* (where bees remove infected pupae from brood cells with developing mites inside), *suppressed mite reproduction* (where the mites fail to produce offspring in the brood cell) and *grooming behavior*, also called *mite biting behavior* (MBB). There is evidence that one or more genetic mutations may be involved as well. This is evolution in action, the direct consequence of exposing a species to this challenging new parasite.

These traits lower mite numbers but do not eliminate them entirely from their hives. This is why I use the term *tolerance* rather than *resistance* when discussing the western honey bee and varroa mites. The bees tolerate a low level of mite feeding without eliminating them entirely.

Unfortunately, mite feeding increases the levels of many viruses, appearing as deformed wings, crawling behavior, shiny black bees, and more. These viruses must be dealt with on a case-by-case basis.

Without a mechanism of proven mite tolerance the continuous maintenance of colonies is a difficult, even when incorporated with methods of Integrated Pest Management (IPM), practices such as drone brood removal, split making and treatment with organic compounds.

This leads to a big questions: why don't we have widespread tolerant stocks in North American colonies? Why haven't beekeepers adapted queen lines that have natural mite tolerance? To be blunt, the beekeeping industry chose to put greater efforts into chemical controls, searching for a silver bullet that would kill these pests. Unfortunately, this lead to mite resistance to chemicals, contaminated combs and honey and a limited support of research into the development of tolerant bee stocks.

Diversity, Uniformity or Both?
A Bit of History of Bee Breeding

Functional Instrumental Insemination

Until the 20th Century, all queens were mated using the system developed by nature inside Drone Congregation Areas (DCAs), where queens and drones mate during a suitable afternoon.

The first person to develop an instrumental insemination device was Dr. Lloyd Watson in 1926. Results were inconsistent until, in the 1940s Dr. Harry Laidlaw Jr., working on his doctoral dissertation on instrumental insemination, discovered the function of a structure called the valvefold. This is located in the queen's median oviduct and acts as a natural barrier to predictable insemination technique. He developed a small device called the valvefold hook, and insemination became easier and much more successful. The use of the hook was simple, moving the valvefold out of the way of a syringe filled with semen. Laidlaw's work was cut short when he was called to serve in World War II. Today inseminators use devices and syringes that even eliminate the need for Laidlaw's valvefold hook.

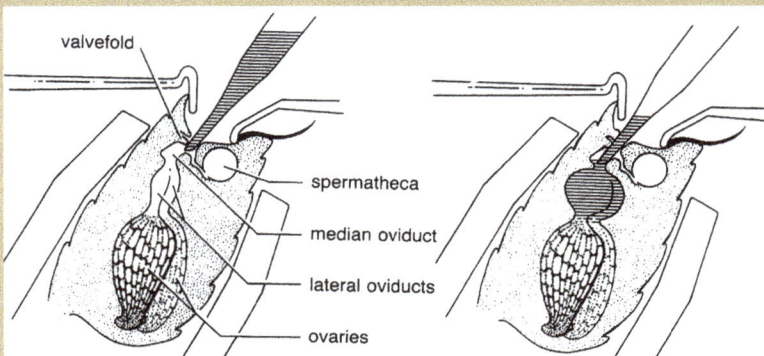

Once Laidlaw discovered the role of the valvefold, the efficiency of instrumental insemination increased. This diagram of the queen's abdomen shows the valvefold being bypassed with a very narrow glass syringe to deliver semen into the median oviduct for proper insemination of the queen. Harbo USDA.

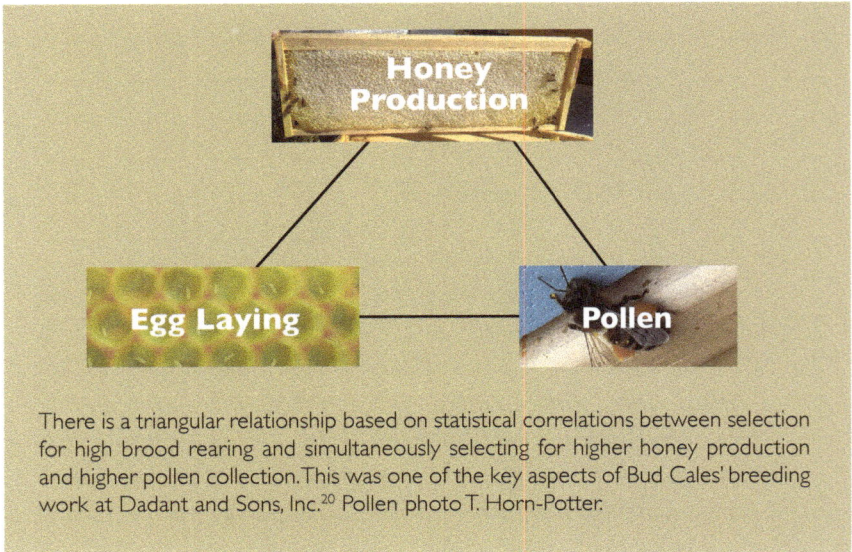

There is a triangular relationship based on statistical correlations between selection for high brood rearing and simultaneously selecting for higher honey production and higher pollen collection. This was one of the key aspects of Bud Cales' breeding work at Dadant and Sons, Inc.[20] Pollen photo T. Horn-Potter.

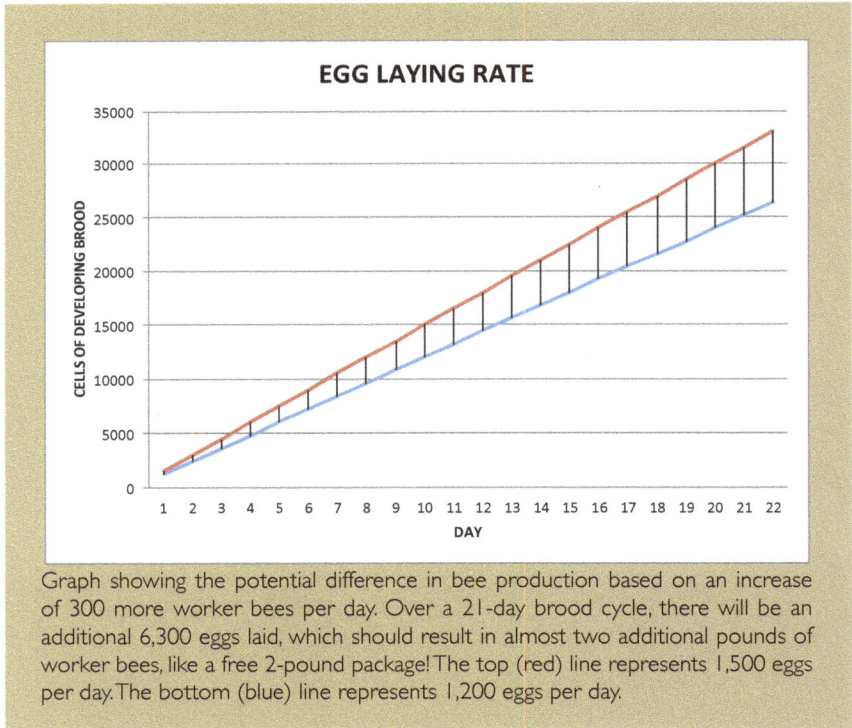

Graph showing the potential difference in bee production based on an increase of 300 more worker bees per day. Over a 21-day brood cycle, there will be an additional 6,300 eggs laid, which should result in almost two additional pounds of worker bees, like a free 2-pound package! The top (red) line represents 1,500 eggs per day. The bottom (blue) line represents 1,200 eggs per day.

Hybrid Honey Bees Modelled After Hybrid Corn

As a much younger man I worked with Dr. G. H. "Bud" Cale of Dadant and Sons, Inc., another veteran of WWII. He was responsible for the development and production of hybrid honey bees, working in Iowa with university hybrid corn breeders. The main breeder queen line was called the Starline Hybrid honey bee.[21] It was a golden bee selected for predictable uniformity, high egg-laying rate with strong brood rearing characteristics, high pollen and honey production and good wintering ability. It was generally a nice bee to work and less likely to sting. Breeder queens were created in the laboratory by instrumentally inseminating virgin queens of one inbred line with brother drones from a second line. These breeder queens were sent to cooperators (commercial beekeepers and queen producers) who grafted and raised daughter queens.

The virgin daughter queen cells were placed into natural mating nuclei in an area saturated with drones from the previous year's breeder queens in Cale's crisscross mating plan. In this plan, the drones were also hybrids, carried the genes of a third and fourth inbred line. The beauty of this system is that is uses last year's breeder queens to produce this year's drones.

The resulting colonies were called four-line hybrid bees because four separate inbred lines configured the bees populating the final production colonies. Because of the genetic control of all four lines, the resulting hives were very uniform. If one colony was ready for its first honey super, it is likely that all the colonies at a particular

BUD CALE'S CRISS-CROSS MATING PLAN

Year	Grafting Mothers	Drone Supply
One	**HxG**	**aExF**
Two	**aExF**	**HxG**
Three	**HxG**	**aExF**

Two different two-inbred lines were crossed to make each year's breeder queens. These were alternated each year to mate with the previous year's drones.

apiary were also ready. Starline Hybrid colonies produced a large brood nest and were popular with beekeepers who produced package bees or increase nuclei (splits). The resulting worker bees demonstrated heterosis, or hybrid vigor, were big in size, and carried more nectar as they foraged.

Bred for uniformity, the Starline traded genetic diversity for predictability. This uniformity allowed certain beekeeping manipulations to be completed during one cycle of hive visits. After inspecting a few hives in one location, determining what needs to be done, and manipulate the rest of the colonies in the same manner. All colony work could be completed during one visit with little need for a return visit to the apiary.

These hybrids were designed to be repurchased every year or two, depending on the management plan used by the beekeeper. The bees produced large colonies, and could easily fill two deep hive bodies, about 85 liters of space, with brood comb, bees and stored food in the brood nest. They also filled an another 85 or more liters of surplus honey. Our goals were bigger bees, bigger brood nests, and bigger honey crops. They were perfect for varroa mite infestations.

Comparison of Nest Volume[22]

Average New York Feral Colony	**47 liters (12.4 gallons)** **10-110 liters** **tall, cylindrical**
One Deep Langstroth	**42 liters (11.1 gallons)** **horizontal and rectangular**
Commercial bees in Double Langstroth Deep Hive	**85 liters (22.2 gallons)** **more box like**
Commercial bees in Four Langstroth Hive Deep Bodies (Summer Honey Production)	**170 liters (44.4 gallons)** **vertical and rectangular**

Over the decades it existed, critics considered the Starline to be a genetic dead end, or at least a genetic bottleneck, lacking the genetic diversity needed to fight a potential unpredictable event. But many beekeepers did not like the daughter queens from hybrid bees and did not repurchase new queens, perhaps because of the expense

An Alternative to Hybrids

Meanwhile, as hybrid bees were being developed and marketed, other beekeepers followed a completely different approach to stock management. Instead of inbreeding, they produced daughter queens from *every colony* in their apiary *every year.* They often did this by raising brood above a double screen or a solid board (with an upper entrance facing away from the regular entrance of the parent hive). Now, separated from the mother queen's pheromones, the bees were stimulated to raise a new queen above the parent hive. The double screen warmed the smaller colony without pheromones from below to interfere with queen production and mating. When frames with young larvae are placed above the colony without a queen, the worker bees build queen cells from

Colonies in the front are walk-away splits made from the four colonies. Queenless, each new colonies will produces its own queen. Or, frames with queen cells in development may be used to set up these new colonies. A. Connor.

the worker brood. Once the new queen was mated and the colony was growing in bee numbers, the beekeeper had a choice. Either move the new colonies with the daughter queens to another apiary, or remove the divider and allow the two queens to fight until only one remained—often the new queen replaced her mother.

When the new colonies were set off and used as an increase colony, some beekeepers call them Walk-Away Splits[22]. They are popular because of their ease of set-up.

The process maintained maximum genetic diversity. Each colony produced a daughter. If you had 500 colonies, in theory you would also have 500 different genetic sources. Whenever there were mating failures, new queens, perhaps from a new bloodline, could be introduced and their drones entered the general drone population.

Compare that to the Starline Hybrid, which had four genetic sources, one from each inbred line. The production of 500 different queens, each carrying the genetic information of their queen, is important by itself, but when we factor in the fact that each queen mates with 14-61 drones, the resulting genetic diversity increased dramatically.

Each drone carries different genetic information that adaption for survival, productivity and disease resistance. Research spear-headed by Dr. David Tarpy at the North Carolina State University has shown that multiple drones, represented as different fathers of worker bees in a colony, improve a wide range of genetic traits that add to the genetic diversity, hive productivity and survival ability.

Double screen with a rear entrance, used to produce a daughter queen and colony over a strong, overwintered colony,

Queens produced by Walk-Away Splits varied dramatically in their genetic background. Some were very productive while others were a disappointment. Some were as gentle as pussycats to work, while others were tigers with a personality disorder. Today, the wise beekeeper replaces the poor producers and red-hot mommas using productive, mite-tolerant and gentle queens acquired from other beekeepers, or queens raised from their best colonies. Some beekeepers tolerate more stings when colonies produce more honey.

In the years following discovery of tracheal and varroa mites, the industry lost the Starline hybrid because there was little financial incentive to select for mite tolerance. It was expensive to maintain inbred lines and went through a series of owners and managers.

The hybrid program was not alone. At the same time we also lost a great deal of our country's honey bee genetic diversity. Since the mites became established in the United States in the 1980s, one researcher reported that the commercial beekeepers in the United States were relying on only 300 queen lines for the nation's blood line because so many other genetic lines had been killed by the mites. This is an over whelming and hopeless feeling. It did stimulate researchers and beekeepers to search for diversity, including the introduction of different mite-tolerant queen lines from Russia, Europe and elsewhere under USDA regulation.

Closed Populations

Drs. Harry Laidlaw, and Robert Page developed a system of combining instrumental insemination while maintaining diversity of the selected stock to develop the concept of closed populations.[24] A population is closed when mating is controlled so that only offspring of selected parents are used as parents for the next generation. Genetic material from outside the breeding program does not enter the population. The program is used by Sue Cobey with the New World Carniolan stock. It is also used with the Russian Bee Breeders in a highly specialized structure involving up to 17 breeder lines.

This is a technique, not a strain, stock, hybrid or queen cluster. For the average beekeeper, it involves the elimination of colonies with undesirable traits—as defined by the beekeeper. They may include poor production, high defensiveness, the presence of brood diseases and an inability to ward off varroa mites.

Green plastic foundation with drone-sized cells. are used for drone production for mating, varroa management, or for honey storage.

Drone larvae and pupae exposed during hive inspection, allowing you a quick check for mites on the pupae.

Strains of Bees Selected for Tolerance Against Varroa Mites

The Biology of Mite Tolerance

Since the introduction of varroa mites into North America, there has been considerable discussion about the mite's feeding on developing pupae and adult bees (on fat bodies, not the hemolymph, as long believed[25]) and the mite's ability to transmit viruses or promote their reproduction. These two factors lead to the widespread death of colonies of European honey bees.

Early on, the response was to treat colonies with miticides that, when used properly, controlled the mites and did not kill the bees. Unfortunately, some of the compounds were used improperly and killed bees, contaminated beeswax comb, and resulted in the mites becoming resistant to the chemicals. This saga has continued in the 30 plus years since the mites were introduced into the United States, with the bee industry shifting to so-called natural poisons, replacing the 'test-tube variety'. These efforts were a result of looking for a lower-priced treatment. Unfortunately, any poison has downfalls in wide spread application.

Beekeepers who once yelled and testified against farmers who misused insecticides became primary abusers of pesticide label laws. They sought out cheaper and often untested formulations of chemicals with the same active ingredients and applied them at whatever rate they considered effective. Thousands of colonies were over-medicated, queens were killed, drones were sterilized and combs were contaminated for years.

Smaller Colonies Swarm More Often

Colonies in bee trees and natural nest sites tend to be much smaller than the Langstroth bee colony. When small colonies, nest volumes of 42 L (11 gallons or 44 quarts) were compared to large colonies with up to 168 L (44 gallons), researchers found that the smaller colonies swarmed more often, had lower varroa levels, had less disease and survived at a higher frequency than the bees in the large colonies. Some consider this a form of natural selection, while others consider it a reflection of the highly adaptive nature of bee colonies.

Selection for Mite Tolerance

Reports and research studies show that honey bees use various method of fighting back against varroa mites. We will now review some of these key areas of parasite combat: lines from surviving queens and intentionally selected for genetic traits.

The Survivor Stocks

Russian

Russian queens are bee lines imported from Russia by USDA. These are the best queens found in areas of Russia where *Apis mellifera* and *Apis cerana* were managed by beekeepers in the same region dating back in the era of the Czar. *A. cerana* is the natural host of the varroa mite, reproducing only on the drone brood.

Nowadays, the lines are maintained with the combined efforts of the Russian Bee Breeders Association (RHBA) members in cooperation with the USDA Bee Laboratory in Baton Rouge. They are not hybrids but bred from a closed breeding population selected for tolerance to varroa and increased honey production.

Russian queens were brought into the United States by the USDA by Dr. Tom Rinderer. The bees had been in close contact with *Apis cerana* colonies for nearly 200 years, allowing the bees to develop natural tolerance mechanisms against the mites. USDA.

As of this writing RHBA has 17 lines of Russian genetic stock.

The Russian honey bee lines were released to the industry for breeding after USDA demonstrated the resulting colonies ability to keep varroa mites at low levels. Colonies will have some varroa mites but at levels that do not require massive treatments of hard chemicals. Russian bee breeders recommend a treatment once every 12 to 24 months using a 'soft-chemical' miticide like formic acid, thymol, essential oils, or HopGuard®. Beekeepers are advised to monitor varroa mite levels and treat hives as needed to prevent problems with viruses even when mite levels are not excessively high.

VSH

The Varroa Sensitive Hygienic, VSH, queen stock is another survivor line. Queens from Louisiana, Michigan and elsewhere were collected after the initial decimation by the varroa mites of the late 1980s and early 1990s. These were queens in untreated colonies that were alive when more than 99% of untreated colonies died. These colonies provided the queens and drones used to produce the VSH stock produced at the Honey Bee Breeding and Physiology Laboratory in Baton Rouge, LA. The sole criteria of selection was that the queens and bees survived when all others died. Other traits were incorporated later, primarily because beekeepers did not like the initial colonies.

John Harbo developed the VSH strain from survivor stocks. In retirement, he and wife Carol continue to upgrade the stock and produce breeder queens. Harbo site.

Saskatraz

The Meadow Ridge Enterprise Ltd. of Saskatoon, SK Canada selects breeder queens after two or three years of evaluation, using natural mating in isolated regions. Queens are evaluated for several traits, especially varroa tolerance (VSH activity), high honey production and wintering ability.[26a] Seven different Saskatraz families are maintained for genetic diversity. Daughter queens are open mated by Olivarez Apiaries in Orland, CA.

Saskatraz colony. Genetic stock bred in northern Canada, the queen produced and mated in California and heading a colony in my back yard in Kalamazoo, Michigan.

Sealed brood being uncapped, evidence of hygienic behavior. File.

Liquid nitrogen being used to kill a sample of healthy brood in the Minnesota Hygenic program to determine sealed brood uncapping and removal rates. U. Minn. Entomology.

Specific Trait Selected Stocks

Minnesota Hygienic Queens:
Bees detect and remove diseased or mite-infected pupae from sealed brood

This strain of hygienic bees was developed at the University of Minnesota by Dr. Marla Spivak from bees at the university apiary, primarily Starline hybrid colonies and their daughter colonies. Spivak selected for a high degree of hygienic behavior effective against varroa mites and diseases of the brood such as American foulbrood and chalk brood. An area of sealed brood was killed with liquid nitrogen. The percentage of cells uncapped and removed by the bees at 24 and 48 hours were recorded. This trait utilizes two behaviors acting in synergy, the uncapping of dead or diseased cells, and the removal of the pathogen or mite, along with the pupae, disrupting the disease and mite life cycle. Spivak supplied the stock to bee breeders to help increase tolerance in the country's honeybee population. There are two suppliers of the stock.

Mite Biting Behavior (MBB):
Bees groom each other to remove and damage mites

As noted, the varroa mite evolved on the Eastern honey bee *Apis cerana* where it only reproduces on drone brood. In addition, the

70

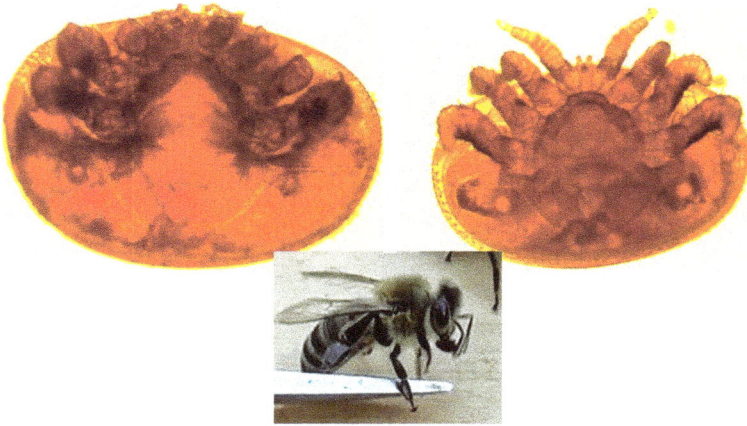

L: Varroa mite that was bitten and legs chewed by a bee from of the Purdue MBB study. R: Live mite without bite damage. D. Wells. Inset: Worker bee biting a mite, from P. Smith video.

Eastern honey bees groom each other, removing varroa mites off the bodies of their sisters and biting the parasites, causing damage or removing their body shell, antennae and legs. This eventually kills the mite via dehydration and exposure to pathogens.

This behavior was utilized by Dr. Greg Hunt from Purdue University to develop the Mite-Biting Behavior (MBB), aka 'Purdue leg-chewers'. Stock has been released to interested queen producers and leaders of bee clubs throughout the Midwest. The Purdue stock makes more honey than the other test groups. The bees have increased winter survival and honey production as well as reducing the cost of replacing dead colonies. The stock also increased profits from honey sales.

Grooming is a trait that is effective against the varroa bomb, a condition that occurs in the late summer or fall as colonies collapse from high mite loads. Once a hive becomes weakened two things happen. It is thought that bees from other hives rob the colony and pick up mites that are carried back to their colony. Also, bees from the infected hive leave the hive covered with mites and enter foreign colonies, spreading the mites. Colonies with low mite numbers, even hygienic lines, can quickly be overcome by mites.

Freshly inseminated and numbered queen from MBB stock. D. Morgan.

African bees deter varroa mites with a faster developmental period and frequent swarming. D. Caron.

Stocks with Developmental Adaptations

Shorter Brood Cycle Bees:
A shorter worker brood cycle reduces mite reproduction.

African and Africanized populations of *Apis mellifera* develop in 19-20 days, compared to 21 days for European bees, providing a selective advantage against mite populations. Unfortunately, faster worker bee development may or may not correlate with shorter drone brood development. This is seen in several African races, including the African bees found in North America (Table 1).

Suppressed Mite Reproduction:
Mites fail to produce daughters

The survivor bees collected by USDA in Baton Rouge were first called SMR (Suppressed Mite Reproduction) mites, because they were found in cells without daughter mites. The only mites left in the cells were nonreproductive or sterile. Initially described as a mechanism of tolerance, this USDA Baton Rouge survivor stock was later renamed the VSH (Varroa Sensitive Hygienic) stock. This happened when it was discovered that the mites with daughters had already been removed from the cells due to hygienic

Table 1.
European vs. African Honey Bee
Developmental Time from Egg to Adult

European	African
Queen...16 days	Queen...14 days
Worker...21 days	Worker...19-20 days
Drone...24 days	Drone...24 days

behavior once they started to reproduce inside the brood cell. The reproduction of mites triggers their removal by the bees. Mites that did not produce daughters were not removed. One of the scientists who work on this project, Dr. John Harbo, now retired from USDA, works with his wife Carol to improve the VSH stock. They believe it has some mechanism of reproductive suppression at work. They continues to produce breeder queens which they maintain and improve.

Gene for Resistance:
Varroa cannot biosynthesis ecdysone

Since the early years of varroa in Europe, American born Dr. John Kefuss has researched natural resistance in his commercial beekeeping operation in Toulouse, France. Maria Bolt now runs the program. Now, decades later, this stock has shown to have a gene for resistance. Here is the abstract from work with their bees.

A gene for resistance to the Varroa mite (Acari) in honey bee (*Apis mellifera*) pupae[28]

"The combination of a virulent parasite and relatively naïve host means that, without acaricides, honey bee colonies typically die within three years of varroa infestation. A consequence of acaricide use has been a reduced selective pressure for the evolution of varroa resistance in honey bee colonies. However, in the past 20 years, several natural selection based breeding programmes have resulted in the evolution of varroa-resistant populations. In these populations, the inhibition of varroa's reproduction is a common trait. Using a high density genome wide association analysis in a Varroa-resistant honey bee population, we identify an ecdysone-induced gene significantly linked to resistance. Ecdysone both initiates metamorphosis in insects and reproduction in varroa. Previously, using a less dense genetic map and a Quantitative Trait Loci analysis, we have identified ecdysone-related genes at resistance loci in an independently evolved resistant population. "Varroa cannot biosynthesise ecdysone but can acquire it from its diet. Using qPCR we are able to link the expression of ecdysone-linked resistance genes to varroa's meals and reproduction. If varroa co-opts pupal compounds to initiate and time its own reproduction, mutations in the host's ecdysone pathway may represent a key selection tool for honey bee resistance and breeding."

There is a second Asian mite to keep out of North America. Here we compare L: *Varroa destructor* with R: *Tropilaelaps* spp. Bee Informed Partnership.

4. OBTAINING BEES

The method beekeepers use to obtain bees impacts colony survival.

Beekeepers new and old have various options for obtaining bees, ranging from catching swarms to buying established colonies. What works in one region for one beekeeper may be a poor choice for someone in another location. Environmental and climate differences often influence beekeeper decisions. The advantages and disadvantages to each method are important to discuss because there are a number of ways that bee colonies can be masterly employed or ineptly misused. In this chapter we will review the pros and cons of buying full-sized colonies compared to nucleus colonies or packages of bees. Also, we will discuss bee removal, swarm catching and using bait hives.

Complete, Full-sized Langstroth Hive

Advantages: Arguably the easiest way to start in beekeeping. Most of the work of building up a colony has been completed, the queen is laying, and the bees should have abundant food to survive.

Disadvantages: Expensive and potentially heavy to lift, especially by one person. Large bee numbers often intimidate new beekeepers, sometimes reducing the frequency of colony inspections, making this is a poor choice. During a shortage of food, large colonies can become very hungry and require sugar feeding to protect the investment.

There are many advantages of starting with complete hives. Complete hives contain adult bees and a full range of brood—the eggs, larvae, pupae. There must be a mated queen in the hive that should be demonstrating her viability by building this colony to a respectable size. The total population of bees should be growing rapidly, and the colony should be strong enough to accept additional boxes of drawn frames, beeswax foundation or frames with starter strips. While the colony grows, the beekeeper's experience and knowledge should also grow.

Complete bee hives consist of eight- or ten-frame boxes of bees and brood, pollen (stored as bee bread) and honey. Most of the comb should be fully drawn—where the bees have built fully formed beeswax comb leaving little undeveloped comb areas. There may be extra boxes included in the purchase, as determined by the seller and buyer coming to an agreement on the number of frames of bees and brood included. There are no industry standards for any bee hive purchase. There may, however, be locally-accepted practices that influence beekeeper behavior.

For example, the size of what beekeepers consider a full-sized colony differs in various parts of the country. In Florida, a full-sized colony is one deep box of brood and bees. But in my Michigan location, a full sized colony has two or even three deep brood chambers (or the equivalent in other depth hive bodies). Some of my Kentucky and Tennessee beekeeping friends keep bees in a brood chamber that consist of one deep and one medium hive body. Southern colonies have a lower egg-laying rate than northern colonies at peak season, influenced by differences in the changing day length and the amount of incoming food provided by nature.

Complete Langstroth hives may be expensive, but at certain times during the season—sometimes as bees come off pollination contracts—they may go on sale. Assuming they are in good shape and you are ready to expand your colony count and produce a

L: Eight-frame; M: ten-frame; and R: five-frame Langstroth hives. All use the same deep frames. A. Connor.

crop of honey, this would be a wise investment. One potential problem with complete Langstroth hives is that they have a risk of carrying diseases and parasites, but this concern is not unique. Many other methods that use previously-owned bees also share this risk factor.

These popular options of a single deep box; a double deep hive; or one deep box and one medium box may simply reflect what the beekeeper selling the hives learned years ago. A ten-frame double box hive often requires two people and a dolly to move because of its size, weight and bulk.

If the combs in a complete colonies are old, then replace them. Some large commercial beekeepers intentionally 'out-cycle' their old combs into full hives and nucleus hives intending to sell them to other beekeepers. This eliminates old, contaminated combs. This is not good for the naive and unsuspecting person attempting to become a sustainable beekeeper. Avoid equipment with black and broken brood combs. These combs may be saturated with pesticides and environmental contaminates, spores that cause bacterial, fungal diseases, and viruses. Some of the comb should

Medium Langstroth frame from the brood nest of the hive. There is worker brood in the center of the frame. There is sealed honey on the upper right- and left-handed corners of the frame, suggesting this colony has plenty of stored honey and may need additional storage space. A. Connor.

be relatively new (light to medium brown color) but the old, black, contaminate combs should be removed as soon as possible.

Seek a hive where the queen was produced in the same season in which the hive is purchased, or one produced after the previous year's summer solstice so it is not too old and potentially about to swarm or supersede. Large colonies without surplus honey may require a great deal of feeding before being ready for the start of winter. When a colony is light on stores, factor in the cost and labor of feeding 50 to 100 pounds of 2:1 (sugar to water ratio) heavy and sticky sugar syrup.

Increase Nucleus Hive

Advantages: An excellent way to start beekeeping. Nucleus hives should have almost all developmental work underway, the queen is young and laying, and the bees should have some stored food. In a month or two, the colony will hopefully compare well with a full-sized colony. Ideal for new beekeepers to grow with, to use as a learning hive. The beekeeper grows in knowledge as the colony grows. Less expensive than full-sized colonies, but be suspicious if the price seems too low. Lower in weight, they are usually easily handled by a single beekeeper.

Disadvantages: Nucleus colonies are only as good as the person who makes them in the apiary. May have old combs. Unmanaged, nuclei colonies rapidly become crowded and swarm.

Five-frame deep nuclei. Split off from the parent hive in mid-spring, these colonies were provided with new queens and fed sugar syrup to grow and develop.

An increase nucleus colony (or hive) is like a complete hive but in minimum. It should contain all the necessary components of a full-sized hive. Usually, there are four or five frames of bees and brood along with some honey and pollen (bee bread). Nucleus hives are called by many names such as splits, divides, nucs, nukes, nooks, set-offs, walk-away-splits and more. I call them increase nuclei as they are an ideal way to start and increase your colony numbers.

One advantage of buying increase nucleus hives is the reduced purchase price when compared to full-sized hives. Nucleus hives ideally contain a younger queen from the same or previous season produced from mite-tolerant bee stocks adapted for your area, or a new queen introduced into the unit about a month or more prior to its sale and delivery to the new owner. Overwintered nuclei

Started in May and June using swarm cells from mite-tolerant queens, and put into five-frame polystyrene boxes, these colonies became support hives over the summer and were allowed to keep all of their honey over the winter.

have the added advantage of surviving a winter in a particular location, which is considered a desirable trait. Spring nuclei made early in the season often contain queens produced and mated in Hawaii, Florida or California and hopefully have been laying eggs for a month or more. The buyer is advised that these are not locally adapted queens.

Should a new beekeeper consider getting a nucleus hive as her or his first colony? My answer is yes. When you look at the three year survey done in northern Virginia (Table 2) you see that the locally produced nucleus, furnished with a mite-tolerant queen, produced the highest overwintering success (87%) of the various options reported. While this study is very limited in scope, it does reflect the experiences of hundreds of beekeepers every year.

The disadvantages of using nucleus hives are the same as found in full-sized colonies as far as disease, old comb and old queens are concerned. Seriously-minded nucleus producers routinely sell nucleus hives with new queens. The price of a nucleus hive is usually higher than packages but less than full-sized colonies. Sunbelt-produced nuclei may or may not contain queens that are adapted to their final (northern state) destination, although

Colony of Africanized bees. The presence of these bees make Sunbelt queens likely to be contaminated with AHB genes. D. Caron.

Table 2.
Colony Survival After One Winter:

Summary of Beekeepers Association of Northern Virginia Survey Results

Locally produced nucs with mite-resistant queens	87%
Locally produced mite-resistant queens	70%
Bee raised queens (emergency, supersedure, swarms)	65%
Beekeeper produced queens	60%
Queens from Georgia	35%
Locally produced nucs with Georgia queens	25%
Packages from Georgia	20%

(3 years, weighted). Source: Dr. Jim Haskell.

a few very progressive nucleus hive producers graft queens from northern-bred breeder queens. They are worth seeking out.

If mated in areas where African bees are present, the queen probably has mated with one or more African drones, so the buyer must expect a few colonies to have more defensive bees than desired. If you experience unacceptable defensiveness, replace the queens or even destroy the entire colony when you obtain increase colonies that contain queens mated in south Florida, south Texas, Arizona or southern California where African drones are present. That includes some northern area where migratory beekeepers have transported colonies from African bee-saturated regions of the country.

During the last few years, the beekeeping industry has shifted to the production and sale of locally produced nucleus hives that contain sustainably raised queens adapted for local conditions. Produced from stock selected in particular areas, this move has resulted in a major increase in survival, better honey production and healthier colonies that carry adaptive traits suitable for the area. The Virginia survey showed the imporance of locally produced nucleus hives. By using locally-produced queens, hives are alive and healthy the next spring.

Package Bees (Artificial Swarms)

Advantages: A popular and acceptable way to start at a reasonable cost. Easy to install with proper instruction and teaching. The low bee numbers should not intimidate a new beekeeper. Packages are great for re-establishing colonies that died over winter.

Disadvantages: Usually a wise investment but bees may start without stored food and will require feeding for most if not all of the first spring and summer. A relatively small unit that requires growth. New beekeepers frequently fail because they are not properly trained on how to use package bee colonies.

Worker bees are sold by the pound and shipped in plastic or wood containers consisting of bees, a feed can of sugar syrup and a caged queen. There is no brood in a package colony. Bees shipped as part of a package are shaken from the brood frames of healthy colonies and paired with a young queen that was raised and

Shaking a package of bees into a hive body. Once the bees were out of the shipping box, additional frames were added to the hive body to fill the space.

mated in another colony. The queen is probably not related to the worker bees she is shipped with and very often not even of the same genetic race. They have only been together since the day the nucleus shipped, so the bees may not have fully accepted their new queen when they arrive (a four-day minimum recommended). In many package bee operations, she may have been laying eggs for one or two days before being caged to ship with the package bees.

For package bee colonies to be successful, the worker bees in the box must become familiar with the queen's scent (her pheromones). If released too early, the worker bees often kill an unfamiliar queen, unintentionally risking colony death rather than accept an unfamiliar queen. The four-day confinement of the queen allows the worker bees to acclimate to the queen. If, when the queen is released, he workers crawl over her body and curl their abdomens toward her, they are attempting to kill the queen in a process called "balling".

The main advantages of using packages are their lower cost and the simplicity of their introduction into the colony. The chance of developing brood diseases is reduced (but not eliminated) because there are no combs, but the lack of brood often leaves package colony bee populations seriously out of balance. There is no prior relationship between the queen and the bees. Without proper

After the bees were shaken out of the shipping cage, the queen, in her cage, was placed between two frames. The can of syrup, if not empty, provides more food. These frames must be better spaced or the bees will build a thin comb. A. Connor.

care and feeding, packages can languish when the weather is cool and food supplies are limited. I have found that, in some years, packages are very successful and produce a crop of honey, but in other years there are widespread queen problems and all package colonies are dead by fall.

Why is there such a variation? Package bee producers are pushed hard to meet the early spring expectations of northern beekeepers. The queens may be produced early when the bees are subjected to poor feeding (nutrition) conditions during queen rearing. This pushes and rushes the mating process, resulting in inadequately mated queens. The drones may also be too young to mate, or stressed during development, being fed poorly and having low sperm counts.

A high number of queens in package colonies never lay very well and are replaced by the supersedure process within weeks of when the package is introduced. Bad weather may mean that the worker bees cannot forage and build wax, resulting in a reduction of the queen's egg-laying unless there are stored food reserves. The workers will die, and without foundation for a queen to lay, she too may be compromised when the bees do not take kindly to her (this is out of the beekeeper's control).

A reduced brood area and the presence of emerged queen cells means the colony has probably just swarmed. There is a slight chance that this is a queen failure and the colony is replacing a defective queen, even a few weeks after installation of a package.

Without established full-sized or support nucleus colonies providing a queen or frame of brood for queen production, package colony with queen failures are usually doomed. In general, package bee users lose about one third of their queens during the first season. In some seasons, the loss rate is much higher. The package bee buyer needs to calculate the risk of losing a percentage of their package colonies, it increases the final cost of the surviving hives.

Comparing packages with nucleus hives, new beekeepers need to adjust the final cost of the packages upward to cover the losses that they may experience using these units. This does not happen every time. I have had very good luck using packages in some years, but in other years neighbor beekeepers have lost 80 to 100% of their package colonies due to queen problems. This makes the final cost of the successful package colonies very expensive. For this reason, I urge new beekeepers avoid package colonies until they understand how to compensate for colony weaknesses and replace defective queens. Keep in mind that my recommendation is in direct opposition with one of the U.S. beekeeping industry's most empowered methods of expanding colony numbers.

Why is this? There is a strong economic incentive by established beekeepers, bee club officers and bee equipment suppliers to arrange for loads of package bees to be delivered into their area to service the needs of their members and customers. It is a simple means of obtaining bees at a low cost to replace losses they have experienced over the previous winter. In most of the United States, package bee colonies out-number nucleus colonies.

Having worked and being friends with package bee producers, I find these people to be hard working beekeepers well aware of the limitations of producing and selling package bees. Most are focused on customer service.

Canadian beekeepers, who cannot import package bees from the United States (but get them from several southern hemisphere sources), rely upon locally produced nucleus hives for spring increase and have been successful doing so.

Captured Swarms

Advantages: A natural and bee-friendly method of obtaining new hives. Beekeepers who seek swarms often put their names and cell phone numbers on call lists at local extension, police, fire and nature centers.

Disadvantages: Capturing swarms is a gamble. The swarm may leave before you arrive to pick them off a tree. Some are high off the ground and nearly impossible to reach. Once captured, swarms require feeding and management for potential diseases. In areas of Africanization, the queen may be carrying defensive genes, and need to be replaced. Many swarms are headed by the old queen in a colony, and often are replaced a few months after the swarm colony is established. Some of the virgin queens in secondary swarms may fail to mate, and the colony is lost.

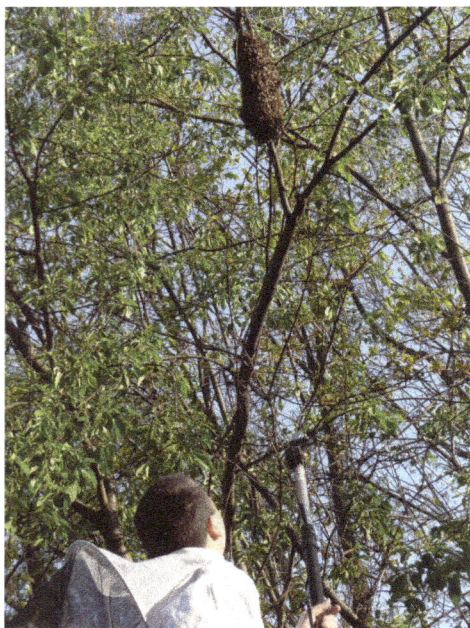

Swarm removal may require special tools, ladders and training in tree work. S. Repasky.

A traditional and respected way to obtain new colonies is by capturing swarms. Swarms generally come from area colonies—perhaps one of your own, though you should not rule out the possibility that a swarm is from a migratory beekeeper's hive. I see that in pollination areas of Michigan.

Swarms are temporary, transient groups of bees that have left the parent hive in search of a permanent location, making one stop along the way. The transient phase may be as short as a few hours or last several days as the swarm sends out scouts to find a new home. Many beekeepers arrive just in time to witness the cluster break as the bees depart for their new home.

Nice sized swarm in a small fruit tree. It may be captured before it finds a new home. T. Ives.

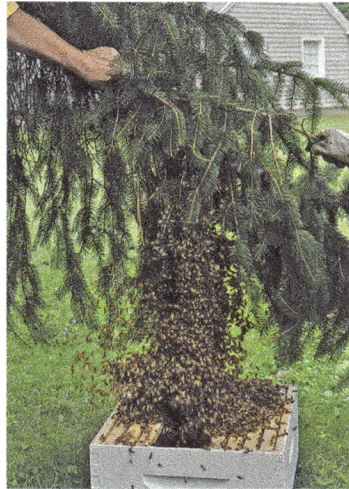

Shaking a low-hanging swarm into empty drawn comb is an excellent way to increase hive numbers.

Swarm season can develop very rapidly in the spring as colonies reach huge brood production from the buildup of population numbers and the abundance of food stored in the hive. Colonies continue to produce swarms as long as they have queens produced for this purpose. Each swarm departs with about half of the remaining adult population. These after-swarms are smaller, contain unmated queens and are less likely to survive.

The first, or primary, swarm from a hive contains 40 to 60% of the colony's adult worker bee population, averaging about 12,000 bees. A prime swarm is an amazing group of bees that often contain more bees than a package colony. The queen within the prime swarm had already built up the parent colony to the point that it can swarm, proving her viability. Within 100 days after a swarm finds a new home, it will have produced 80% of the honey comb it will need the first year. The colony simultaneously focuses on collecting and hoarding food to survive the winter.

Swarm colonies face difficult survival challenges—only one in six survives to become one year old. Fortunately, a skilled, sustainable

A swarm of Africanized bees in south Texas. Colonies swarm frequently and invade colonies with European queens.

beekeeper can increase the survival of swarms by feeding the colony and adding a frame of emerging worker brood to the colony. This increases the number of bees and stabilizes the colony's growth by balancing the hive's work force before the queen's own daughters emerge.

While swarms are thought of as cost-free, consider an accountant's perspective. Add the total of your travel time, lost wages, equipment, feed and other expenses. If you miss four hours of work to capture a swarm and have travelled 100 miles round trip to get to it, a swarm represents a sizeable investment. And, as said above, you also run the risk of arriving at the site only to learn that the swarm has already moved to a new, permanent home. Beekeepers harvesting swarms are often working without any medical, legal or liability protection. A fall off your ladder could be ruinous.

As part of your plan to grow your apiary, consider the capture of a few swarms every year. Put your name, phone number and email on a local swarm list run by a bee club, cooperative extension office, police or fire departments and city hall. Make sure they know how to reach you in a hurry by cell phone should a swarm appear. Make up swarming-catching kits and keep it in all vehicles.

Swarm catching is a great activity for the underemployed, the retired and those with flexible schedules (including college students). Not everyone finds swarm catching a good fit for them, but the beekeepers I know who provide this service do it by being well established as a trusted person to contact directly, rather than responding to Internet postings, where a flood of beekeepers may show up for the easy swarms only to find someone got there first.

Before they departed from the parent hive, swarming bees engorge themselves with honey that sometimes contain American foulbrood disease spores. Some beekeepers medicate swarms with antibiotics as a prophylactic treatment against American foulbrood IF they have the necessary permission from a veterinarian. Consider the new viral phage that kills the bacillus causing AFB.

Otherwise, as a sustainable beekeeper, I recommend you put hives obtained from swarms in an isolated apiary for several months and hope that any AFB spores carried by the bees in their honey stomachs will be killed as they are digested within a few hours of arrival as part of wax generation.

Combine small swarms with medium sized colonies rather than fussing with them all season. Otherwise they may die in the winter. As a general rule, colonies that are weak and starving in the summer will die during the winter, regardless of how much food you provide.

Bee Lining

Advantages: Learn an ancient method of finding bees in remote areas where managed colonies are not present, and colonies are widely dispersed. Continue a tradition kept by a few beekeepers for the good of all.

Disadvantages: Few isolated sites remain, and the bees you 'line' may be owned by other beekeepers. In most areas, bees in bee trees cannot be removed or the tree chopped down.

Special box used in bee lining. T. Seeley.

Bee lining is an ancient bee colony finding technique used to find bees in the forests, popularized in contemporary beekeeping by Tom Seeley at Cornell University.[29] By collecting foragers on flowers, he releases them one at a time and follows the direction they take to eventually find their hive. Bee lining is recommended as a group activity for motivated beekeepers, students, researchers and those accepting the frustration of the process. This is a great activity for those who want to spend a nice day in the fields and woods, ready for a mad chase behind a bee making a bee-line through field and thicket.

Bee Removal (Cut-Outs)

Advantages: A useful craft of removing bees from buildings, other human structures, trees, caves and brick walls. A challenge that is rewarded by potentially feral survivor stocks that are not necessarily like your own. Sometimes there is a great deal of 'wild' honey for the taking, a real bonus. Folks will pay big money to have you remove bees from areas frequented by humans and small, nervous dogs.

Disadvantages: Requires hard work and determination, a tolerance for stings and a specialized knowledge of structure construction (and a lot of tools). Combs must be carefully tied into frames so the bees can rebuild them into movable frames. If care is taken, you will find the queen wearing the bright mark on her thorax. Welcome home Queenie!

Removing bees out of a building, a natural structure or a bee tree is another way some beekeepers increase the number of colonies they own. They can also obtain a large mass of honey. Collecting bees from buildings brings the risk of stings and is a hot and sticky chore. Combs of brood need to be tied to empty frames and the honey combs squeezed out. There is a risk of brood disease, as there is with any unknown hive, and this manipulation can spread disease through robbing by other colonies during the removal process.

There is a unique risk to beekeepers performing cut-outs—property owners often use insecticides in their preliminary attempt to kill the bees before they realized they needed to call for help. By the time a beekeeper is called to remove the bees, the honey and comb may already have been contaminated with an insecticide. What makes this worse is that property owners fail to disclose the truth about using insecticides for fear of losing the beekeeper's bee-removal service. Do not eat honey our use it for bee feed if there is any chance of chemical contamination. Burial is a suitable method of disposal, as honey put into a landfill could be robbed out by other bees.

Colony removal from a wall space. The thicker comb at the top is honey and brood is below. C. Hubbard.

Bait Hives

Advantages: Bait hives are special containers filled with old comb, swarm lures (pheromones) or essential oils that attract scout bees that recruit their swarm to them. Using bait hives to capture swarms is a wise strategy in areas where there is a high density of bee colonies. The bees can easily be moved to standard beekeeping equipment, checked for disease and evaluated for mite tolerance. In areas of the country where African bees are common, these bait hives can be set up in high risk areas (school yards, parks, retirement communities) and the swarms removed and either requeened or destroyed. For a service fee.

Disadvantages: Consider this a business and get insurance. The process is work and the rewards are uncertain. It may require local and state certification and training.

Setting out bait hives is a perfect way to capture natural swarms for growth. In areas of heavy African bee populations, trap nests may be used to remove African colonies and requeen or destroy them as part of an African honey bee management plan. In areas

Bait hive containing old comb and odor attractants get the attention of scout bees. This swarm is in the process of moving into the colony. S. Repasky.

lacking African bees, bait hives are a suitable way to increase colony numbers— and the bees came to you! The disease risk rate of using bait hives are about the same as using swarms.

Bait hives are used to both detect and control African bees in Florida, Texas, Arizona, California and other areas. In the Caribbean Islands and other locations with seaports, bait hives are used to detect swarms that arrive on ships. They are used to monitor and genetically sampling local bees populations, It is one way to obtain naturally-selected traits such as grooming behavior in areas where feral colonies operate without a great deal of intrusion from managed beekeeping practices. The use of essential oils like lemon grass and other lures increase the attractiveness of the cavity. There are commercial mixtures available that may be obtained for this purpose.

Conclusion

Establishing hives or adding new colonies to your apiary offers you many rewards, the confidence of a bit more genetic diversity, and the joy of keeping bees alive. Reviewing your abilities and interests will help you decide which types of hives you seek to obtain.

5. SUSTAINABLE BIOLOGY

Biology-Centric Management

Managing bee colonies with the intention of keeping them alive is a key objective of all beekeepers. Yet they often fail at this task due their lack of understanding key aspects of bee biology. Here are some common examples of how I have seen beekeepers unintentionally, kill bees. A few I have done myself!

A. Splitting a hive during cold weather or when the new hive does not have an adequate amount of bees, chilling and killing the brood.

B. Killing or injuring the queen during a hive inspection.

C. Removing honey at harvest time and failing to feed sugar syrup to compensate for any dearth period (no incoming pollen and nectar) or winter preparation.

D. Temporarily moving colonies for a pollination site or nectar collection yard and not moving them out before the grower applies pesticide compounds known to kill honey bees or interfere with their behavioral development.

Chilled brood: Too few nurse bees to feed and warm these larvae. R. Williamson.

E. Using contaminated hive tools, smokers, bee suits and other bee equipment, contaminating hives with diseases and viruses;

F. Allowing strong colonies to rob out other colonies infested with varroa mites and then dying themselves.

G. Growing a beekeeping operation too fast; creating a situation where there are more colonies to manage than time and energy allow. Colonies may become crowded without honey storage space and swarm or they are close to starvation as winter approaches. If mite levels have exploded, colonies die in large numbers.

Unique Properties of Honey Bees

Physiology

Honey bees evolved in the tropics as social insects. As they migrated into colder climates, they developed methods of heating areas within the brood nest using cellular metabolism and respiration. By eating honey and allowing it to digest, bees generate heat, water vapor, and carbon dioxide. These products are managed by the bees.

Nurse bee feeding a developing larva on a brood frame. Thousands of such feedings occur within a heathy hive on a moment to moment basis.

Beekeeper often cause problems by over insulating and restricting upward ventilation, especially in the winter. Cellular metabolism is the chemical process that occurs within a living organism in order to maintain life. There are two kinds of metabolism: constructive and destructive. Constructive metabolism forms proteins, carbohydrates, and fats that form tissue and stores energy. Destructive metabolism breaks down complex substances like honey into energy and waste.

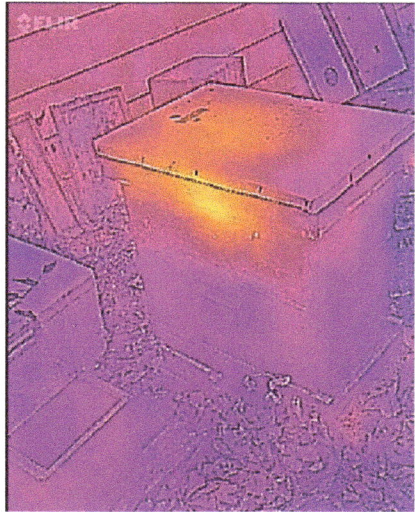

By eating honey, wintering colonies generate heat and are able to raise brood. Thermal image shows this winter cluster high in the hive, suggesting they may soon run out of food. T. Smith.

Cellular respiration helps bees establish and maintain homeostasis (internal stability) to adjust for environmental changes. As a social unit, thousands of honey bees operate as a single, well-coordinated organism, maintaining temperature, humidity and gas (oxygen and CO_2) levels by respiratory heating and physical ventilation of the nest.

Honey bees constantly share food, and most bees are consuming the same food at the same time because of the colony's community stomach. The food sharing process is achieved via trophallaxis, the exchange of regurgitated liquids between adult social bees. The food exchange from bee to bee is continuous. Years ago this was demonstrated by feeding a few bees with sugar syrup containing radioactive isotopes. Within a few days almost all the bees in the hive were radioactive.

There are exceptions, as special feeding rates are not identical when certain bees perform special duties. Some include the nurse bees who feed heavily on bee bread to produce brood food to feed the new generation of developing larvae; wax producer workers that feed heavily on nectar and honey to secrete wax scales that form beeswax; and the queen, who is fed a nutrient-rich diet of to support egg production. Drones require rapid feeding before and during mating, returning every 25 minutes to replenish honey supplies needed to sustain queen searching behavior.

Reproduction: Drones and Queens

Production of the sexual members or reproductive honey bees of the hive, the queens and drones, is seasonally limited. They are normally produced when colony growth stimulates their production as aspects of reproductive swarming.

Queens start as eggs in queen cells. They are either grown in special cups that extend into the bee space of hive frames, or they are produced from ordinary worker cells that are rebuilt for the queen's needs. The larvae receive a food called royal jelly that is fed throughout the larval development period of all queens. Swarm queens are produced when a colony is preparing to swarm, but when an old queen is failing, a new queen is produced through a process called supersedure. If the old queen is accidentally killed, the bees use an emergency response to produce a replacement.

Drones are the overlooked heroes of colony diversity and strength. Every nice afternoon they perform 4 to 6 marathon flights to find an unmated queen. Once mature, they care for themselves and stay out of the way of the workers in the honey combs where they feed and restore their reserves. When they mate, they die.

Mature drone feeding himself from a cell of honey. Both very young adult workers and drones require worker feeding, but feed themselves when older.

Drone pupa with purple eyes above and drone larva below.

Dry protein feeding improves the production of workers, queens and drones in Hawaii. This bag of bee feed was put into an open barrel and cut open.

Keep three frames of sealed ripened honey in each hive at all times to ensure colony nutrition. This is a medium frame. Deep frames hold more food.

Bees produce drones when the colony has abundant brood, bee population and reserves of food. When I was attempting to mass produce drones for use in instrumental insemination in south Florida, I painfully learned that colonies without abundant stored food were highly unlikely to produce drones. By giving colonies a minimum of three or more frames of stored honey and feeding the bees sugar syrup and protein patties, we were more successful in stimulating the production of drones during the early season or when there was no flow underway.

Drone production is also strongly influenced by the change in day length: colonies produce more drones when day length is increasing. Fewer or no drones are produced when the length of the day decreases after mid-October.

Queens mate with many drones during a single or multiple mating flights made 7 to 12 days after their emergence. Sperm from multiple drones are all stored in a structure called the spermatheca, located at the tip of the abdomen, and connected to the oviduct where eggs pass out of the queen's body. The spermathecal duct allows sperm to fertilize each egg as it passes down the oviduct.

After mating, the queen starts to produce hormones that stimulate egg (ova) production, a process called oogenesis. Egg laying is not the queen's decision but one made collectively by workers. Worker bees control the egg-laying rate by adjusting the amount of food they provide the queen, more in the spring and buildup periods and less when the food supply is reduced. Queens will lay from 1,000 to 2,000 eggs per day during the season. My best experience has been with queens that lay about 1,500 eggs per day, or one egg every 58 seconds. 1,500 eggs develop into about one-third of a pound of adult worker bees.

Because of the queen's mating with multiple drones, the resulting colony is considered a superorganism. The hive contains daughter worker bees with different fathers. While there may be 30,000-50,000 or more worker bees in a colony, they represent 14 to 60 different subgroups called super-sisters—each group of workers sharing the same drone father. There is some evidence that super-sisters of the queen may influence certain behaviors as a function of kin selection (favoring one blood line over another), but the overall result of the research is that the colony is focused

Queen laying an egg in a worker cell, surrounded by a retinue of workers.

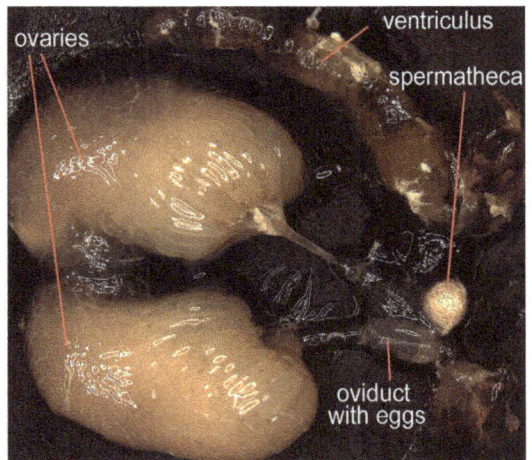

Queen dissected, showing two large ovarioles containing developing eggs and spermatheca. USDA.

Three-year old queen with daughters of different color patterns, reflecting their different drone fathers. Those alike are called supersisters. T. Ives

on overall growth and prosperity, not the survival of one drone's bloodline.

Super-sisters provide additional specialization for certain tasks: water gathering, temperature regulation, pollen foraging, disease tolerance/resistance and much more. This observation supports the need for wide genetic diversity within each bee colony and the benefit of colonies with queens that have mated with a genetically diverse set of drones.

In the wild, where bee colonies are usually sparsely spaced with about one colony per square mile in northern woodlands. In areas of intensive beekeeping, however, there may be several thousand colonies in the equivalent area, providing a large number of drones searching for unmated queens. This sexual competition to mate leads to a greater diversity of drones, thereby improving the success rate of the resulting colonies.

Drone diversity also helps bees withstand many disease infections. Multiple drone mating provides a vast array of genetic information present within the hive, stored as sperm in the queen's spermatheca. This structure stores 4 to 8 million sperm from all these drones. As each drone's sperm is stored in this tiny sac of nutrient-rich material, the sperm become mixed and one drone's sperm fertilizes only a portion of the eggs that become worker bees. Thus, these two mechanisms—multiple mating and drone diversity—serve the bee colony well by increasing the genetic diversity in ways that are far more powerful than a single egg/single sperm union.

The honey bee genome (set of genes each bee possesses) is rich in genes associated with odor and smell, but it has relatively fewer genes associated with taste and immunity functions, reflecting evolutionary adaptations associated with their unique lifestyle. Multiple mating provides a diverse bee genome in living bees that leads the colony's improved ability to fight infections like American foulbrood disease or parasites like varroa mites, as well as a greater number of different bee behaviors.

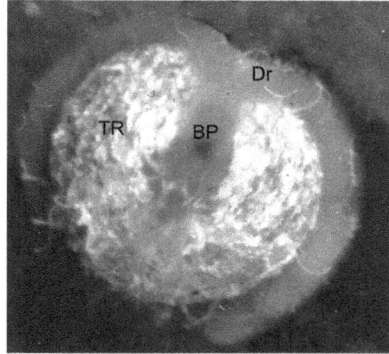

The spermatheca is located at the tip of the queen's abdomen. It can contain 4-8 million sperm. It is covered with tracheal tubing (TR) and a gland (BP). G. Koeniger.

Reproduction By Swarming

Bees, like other insects, produce males and females that use copulatory sex as a means of reproduction. But, as a social insect, honey bees also reproduce by dividing their entire colony into multiple parts in a process called swarming. When a colony gathers large quantities of food supplies and increases its population as a result, the colony swarms. They split the into a new hive, issuing a swarm: about 12,000 workers, a few hundred drones and one or more queens from the parent colony.

When a swarm finds a new home and establishes a separate colony, it ends further contact with the parent superorganism. Parent colonies may issue swarms more than once during each swarming cycle; secondary and tertiary swarms and even more as the population of bees allows. Some colonies produce swarms more than one time a year; a second swarming season is timed with the late summer flowers like goldenrod and aster bloom.

Not every colony swarms every year. It is generally thought that younger queens are less likely to swarm than older queens (or their respective colonies), and strong colonies have a greater probability of swarming than weak ones, regardless of the age of the queen heading the parent colony.

Certain races of *Apis mellifera* swarm frequently—as often as every month. This occurs with certain African races of honey bees (*A.m. scutellata*) that are called African or Africanized bees found in southern regions of the United States. As a tropical sub-species, these bees use frequent swarming as a strategy to increase their chance of survival. Rather than hoarding food, these bees convert incoming food into bees. Sub-species (races) that evolved

Likely a prime swarm because of its size. S. Repasky.

in temperate regions of the world store surplus pollen as bee bread and nectar as honey and keep it available for winter survival or any period of dearth, when natural food is absent.

Nutrition and Growth

The frequency of swarming is directly associated with the abundance of pollen and nectar entering the hive. Large amounts of nectar and pollen trigger greater worker and drone production[30] and eventually the appearance of swarm cells.

Every region of the world has a peak period when swarms are abundant. It may be in February in Florida and May in Massachusetts, reflecting developmental differences in the season, temperature: colony growth and abundance of food at these locations. Where there is a late summer or fall swarming season, bees have experienced increased food abundance. Fall swarms must grow quickly to survive winter, and many are not successful.

Pollen is packed at the flower into the corbiculae (a bee's two pollen baskets) and carried back into the hive. It is placed into a cell of the honey comb located next to the brood area. Bees pack pollen with the flat surface of the bee's head along with the addition of tiny amounts of honey from the stomach of the worker. The honey contains microbes (fungi and bacteria) that lead to a process called lactose fermentation. This lowers the pH of the pollen, making it more acidic, which preserves the proteins and nutrients in the pollen.[31] The food value of dry pollen is lost quickly.

When we compare bee bread with dry bee pollen, we find a higher protein quality in the bee bread, with some proteins predigested into amino acids, making their absorption easier. This improved digestibility makes bee bread so valuable to the bees.

On a frame held by Brandon Pollard, we see brood expanding as the season progresses. The comb shows the initial center size of the brood area, represented by the dark brood in the center of the comb. A second wave of egg-laying is reflected in the lighter colored brood on the outer circle, when nectar flowed.

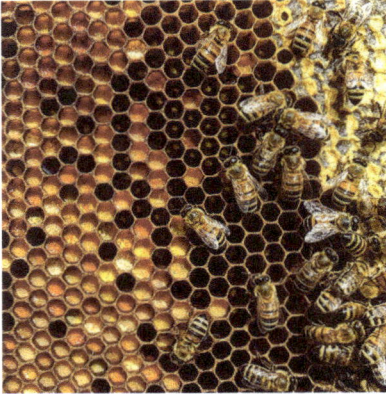

Worker cells containing bee bread (fermented pollen) directly contributes to royal jelly production used to feed all young brood. S. Way.

Newly collected nectar has been converted to honey, providing carbohydrate energy for bees as simple sugars.

Bee bread contains more vitamins, increased vitamin K, more antioxidant concentrations and enzyme levels. Minerals are freed from the cellulose portion of the pollen, including zinc, magnesium and silica. Bee bread is an energy-rich food that increases the life of the pollen, keeping it viable for over a year, compared to few days for dry, unprocessed bee pollen.

Nectar conversion chemically changes floral nectar, which starts as 20-40% sucrose, into a thicker solution made up of two simple sugars, glucose and fructose, which are formed when the bees add to nectar an enzyme called invertase. The process starts with the bee on the flower. When back at the hive, house bees working as honey processors receive the honey and expose tiny droplets of nectar with their mouthparts to the warm air in the hive and allow for rapid dehydration. The resulting 'ripe' honey contains less than 18% moisture and mostly glucose and fructose. The lower moisture level and acid content maintains honey nutritionally stable for years.

Liquid honey after extraction.

SPECIALIZED ANATOMY OF THE HONEY BEE

Along with the social behaviors of bees of foraging, feeding, swarming and queen replacement, honey bees have evolved a wide range of physical and biochemical attributes contributing to their survival and success. Here is a partial list of some of these features:

Forked or Branched Hairs

Honey bee workers are covered with plumose hairs. These fine hairs are forked or branched to collect pollen as the bees work flowers. These hairs are groomed by the front legs and moved to the pollen baskets located on the hind legs. Even the eyes of bees are covered with branched hairs, allowing the removal of pollen. The body of the bee, and thus these hairs, develops a positive charge as the bee flies through the air that attracts negatively charged pollen found on flowers.

These hairs are easily affected by soap-like compounds called surfactants. Used extensively in pesticide sprays to help the chemicals stick to the plant, surfactants also 'wet' down the bee's branched hairs and eliminate the bee's ability to forage. When you see a forager that looks wet on a rainless day, there is a good chance she has been in contact with an agricultural or landscape pesticide application. Bees cannot easily clean this material off their bodies and rely on grooming by their sisters. If associated with the wrong insecticide, the bee will probably soon die.

Pesticide-contaminated pollen stored in pollen baskets can impact colony survival. When bees collect pollen contaminated with pesticides, especially the kind that do not kill them immediately, the bees will store it in the pollen cells. If they eat the poison, it may affect development, life-span or cause death. If the compounds have an unusual odor or repellency, the bees may coat the cells with propolis, a bee-collected mixture of tree resin and beeswax.

Antennae Cleaners

These are specialized structures located on the front legs of workers, queens and drones that allow bees to wipe pollen and other debris off their antennae. Any dirt, pesticide residue and materials of the size and shape of pollen are likely to be groomed from the head and antennae. These and the pollen will end up in the pollen pellet.

Pollen Baskets

This is the pollen storage structure used during foraging, located on the hind legs of workers but absent on queens and drones as they do not forage. These are a complicated set of combs and rakes that allow the bee to pack pollen in front of the flower while hovering, grooming herself with her legs and antennae. Bees also pack pollen while standing on the flowers. The pollen is moved to the hind legs and raked into the corbiculae, a plate-like area of the hind leg that is surrounded by long hairs or spines. The bees visit many flowers to fill these baskets, although a few flowers produce enough pollen to fill the baskets in one flower visit, such as the night-flowering *Cereus* cacti.

Worker bee showing hairs on her head, two antennae, and the sucking straw mouthpart with a tiny, hairy tip to pick up tiny amounts of liquid.

Once the baskets are full (sensed either by volume, weight, or both), the bee returns to the region surrounding the brood nest in the hive. The bee finds cells partially filled with pollen, or empty cells next to them, and backs her hind legs into one cell and reverses the packing motion, removing the pollen pellets from a fine spine in the corbiculae to remove the pellet. These pellets are left for house bees to process, packing the material with their heads, adding honey-containing microbe-nutrients and allowing the pollen to convert into bee bread.

When agricultural chemicals are the size of pollen grains, or if the chemical (dust or liquid) covers the pollen, it is collected and placed in the cells with the pollen. Some insecticides immediately kill any bees that come in contact with the poison. Less potent chemicals mix with the pollen and eventually interfere with bee development. The agricultural chemicals affect the development of the larvae and behavior of the adult bees. Some of these compounds have been linked to poor sperm production, migration and viability in drones.

While they do not carry a warning on the label to protect bees, fungicides (disease-killing pesticides) have been shown to affect honey bee behavior and survivability. Commercial beekeepers hear the roar of spray equipment applying fungicides to almonds while the flowers are in full bloom and the bees are in full flight. The beekeepers explain that the spraying does not affect their bees. Yet, in my personal experience, when the ground is sprinkled with dead and dying bees at the density of one more bees per square foot, bees are being killed. Of course, when you bring bees into almonds for nearly $200 per hive, you are probably willing to accept some bee loss. But now pollinating beekeepers are starting to recognize the negative impact of these compounds on their bee colonies growth and development for the rest of their season.[43]

Hind leg with pollen basket (corbicula). A. Connor.

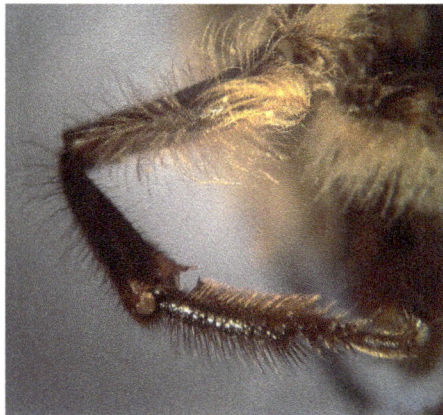

Antenna cleaner on the front leg of a worker. Note body hairs. A. Connor

Forager on *Centaurea cyanus* (cornflower or bachelor's button) with off-white pollen in her pollen basket (corbicula). The bee is also gathering nectar.

Large Eyes

Inside a dark bee hive many complicated colony communications take place through pheromones, odors, touch and vibration. Outside the hive, bees use their amazing vision to complete their tasks. Worker bees have the smallest compound eyes, filled with many hexagon-shaped facets that focus images that are excellent in detecting motion and color. Bees do not respond to the wavelength of red, but can see ultraviolet, a wavelength visible to humans using special cameras and films that show what a bee may see.

Drones and queens have larger eyes than workers, and these are essential during mating. Since drones must find queens, and not the other way around, the drones have the largest eyes to detect and follow queens in the air, hundreds of feet off the ground.

Honey Stomach

Worker bees have a very specialized stomach. In cows, which have four stomachs, the compartments contain food at different stages of digestion. With bees, there is a special section of the digestive tract that is separate from the rest of the bee's gut system. This

Drones come with large eyes, large antennae and a larger body size. USGS Bee Inventory and Monitoring Lab.

stomach is a storage sac, or honey stomach. The enzyme invertase is added to the nectar as it is collected by foragers, and this breaks the sucrose compound into glucose and fructose.

A special valve separates the honey sac from the rest of the gut of the bee. This valve is called the proventricular valve, and it allows the passage of solid food, mostly pollen, into the gut but not the liquid portion. Thus, nectar is not vomit, but regurgitated nectar with much of its pollen removed.

The important part is that the nectar is not digested any further but held in a more stable but partially modified stage. When the bee returns to the hive she goes to a section of the hive where honey is being stored. This may be in the brood area or in the storage comb area above the brood nest. She does not deposit the nectar directly into a cell, but the liquid is shared with one or more bees whose job is to ripen the nectar into honey. This is done by the regurgitation by the field bee and the uptake of the house bee.

Nurse bees consume the stored honey and bee bread and digest it fully, providing the energy and nutrients needed to feed the queen and the growing larvae. They also feed young workers and drones before they are able to feed themselves.

Forager (R) offering nectar to a house bee (L) who then exposes honey bubbles to the air to reduce the moisture and finishing honey conversion. R. Williamson.

Sting and Venom

As painful as bee venom may be to a human, you must appreciate that the sting is a pretty remarkable structure. Over the course of evolutionary time, the sting structure was modified from the ovipositor of primitive wasps into two barbed shafts that slide beside each other to puncture tissue. Evolved to defend colonies from animal intruders, it works fine on humans, leaving the sting structure in the tissue. While there is little pain from the sting itself, it is the venom that provides the defensive reaction through painful learning and complex physiological effects on the target.

Worker bees and queens have sting structures. Drones do not. Bee venom is made of a complex group of proteins, a mixture of 60 chemicals. Some are anti-inflammatory, including melittin and adolapin. Melittin is the main component (40–60% of the dry weight) and the major pain producing substance of honey bee venom. Melittin is a basic peptide consisting of 26 amino acids. Some compounds increase nerve transmission, including Apamin, and the neurotransmitters dopamine, norepinephrine and serotonin. The inflammatory response after a sting is the work of hyaluronidase, phospholipase A2, histamine and mast cell degranulating protein (MSDP). The venom of honey bees is different from other stinging Hymenoptera, so protection from one venom does not guarantee protection from the venom of another bee or wasp species.

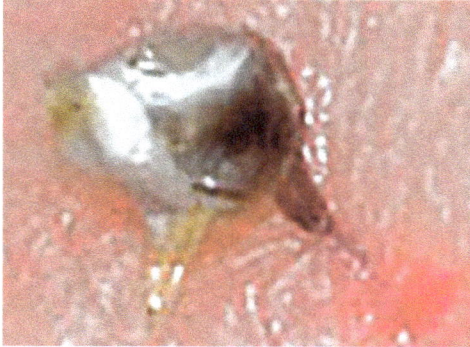

Structure of worker bee sting allows continuous pumping after stinging.

Mouthparts

Consider having two sets of mouthparts like honey bees: the mandibles and the proboscis. Mandibles are solid structures bees use for chewing and shaping. They move back and forth and are used by emerging adult bees to cut their way out of pupal cappings, then later to shape beeswax and bite varroa mites. The proboscis is a foldable, straw-like structure, smooth and tube-like but made up of modified mouth structures allowing the bee to collect tiny amounts of liquid, such as nectar. At the end of the tube is a tiny sponge-like structure that soaks up the last bits of nectar and contains taste sensors. The proboscis is used by bees to exchange food in a process called trophallaxis. Food exchange is part of honey making as nectar passes from bee to bee.

Mouthparts (proboscis), antennae and hairy body of worker bee. USGS Bee Inventory and Monitoring Lab.

111

Drones have very short mouthparts but the largest eyes and antennae of the bees. USGS Bee Inventory and Monitoring Lab.

Antennae

There are two, long and segmented sensory filaments on the front of the head of a bee. Often called "feelers", they are attached to the head with a ball and socket structure that allows full rotation. The base is called the scape, which has two muscles that swing the antennae up and down like a hinge joint. This leads to the pedicel, which has two more muscles that move up and down. The pedicel contains the organ of Johnston, a sensory organ that detects sound vibrations. The final part of the antenna is the flagellum, which is covered with thousands of sensilla or sense organs. Drones have the greatest number of sensilla allowing them find queen pheromone in Drone Congregation Areas (DCAs) where mating occurs.

Genetics and Inheritance

Characteristics of honey bees are passed from one generation to the next through genes located on the DNA of the bee. The genome of the honey bee has been mapped and is documented. Unlike mammals, sex determination of bees is a function of the number and type of alleles for sex. Drones have only one type of sex determination allele, while females have two different types or alleles. Most female larvae become workers, while a few female workers are fed a special diet when the bees decide that they are the ones to become queens. This change appears to be determined by the food the future queen larva receives during her life, and is an example of epigenetics, the study of changes in organisms caused by modification of gene expression rather than a variation of the genetic code.

Some behaviors of bees are hard-wired into their genetic code of the bee, while other behaviors are influenced by learning. Bees apparently understand some basic aspects of mathematics. They can tell time as evidenced by their returning to a plant just as it is about to produce pollen or nectar. Feeding studies have documented this in great detail, where feed is presented to the bees at the same time every day. If the researchers miss a day, the bees will be there waiting, and continue to visit for several days at the same time, with the memory fades in several days. Bees are learn, do math, tell time, and stop previously performed acts.

Worker bee licking the queen to obtain queen pheromone. These retinue worker bees move about the colony and 'spread' the chemical information. R. Williamson.

Pheromone Production

Queens, drones and workers produce a wide range of pheromones— chemicals stimulating a response in a member of the same genus.

Queen pheromone is made of a complex chemical mix of compounds that regulate colony stability, suppress egg-laying by workers, provide swarm cohesion during swarming, and attract drones during mating, but not inside the hive.

Worker bees fanning their wings and some are scenting, exposing the gland at the top of their abdomen, the scent or Nasanov gland. A. Connor.

Workers use other pheromones to work together, mark flowers they visit, call to other bees when the colony is disturbed, and cause other bees to sting at the same location where another bee has just stung.

Drones have a signature chemical called drone brood pheromone. The ectoparasite *Varroa destructor* uses this compound to move into drone cells just before the cell is sealed. It is a kairomones, working between individuals of *different* species while a pheromone works between individuals of the *same* species;

This chemical relationship between varroa mites and honey bees could someday help develop a very targeted mite control method, perhaps trapping the mites. For now, beekeeper can remove sealed drone brood before the drones emerge to reduce the varroa mite load without using chemicals.

We have much to learn about all the colony's pheromones as they have a very strong link to colony survival. Bees that live with their body chemistry in balance produce stronger colonies and are less likely to die.

Drone numbers are controlled by worker bees, not the queen. Drone production is usually associated with a richness of forage entering the hive, but is also influenced by increasing photoperiod. Numbers may be 'trimmed' when in excess, or eliminated completely by worker bees, a useful event to observe and help you guide further management decisions.

Conclusion

The special physical and behavioral characteristics of honey bees allow much greater flexibility in dealing with change, environmental contamination or anything that interrupts holistic functioning of the colony and reduces the daily function of both the bee and the colony.

Worker bees attempting to reject or prevent entry of a drone. C. Abromson

6. SUSTAINABLE MANAGEMENT

An overview of sustainable management

There are a lot of opinions about the meaning of sustainable colony management. Let's summarize and simplify key aspects of this subject. First, there are two management concepts all beekeepers need to pro-actively adapt to maintain and grow their colony numbers:

1.   Varroa mite tolerant stocks
2.   Produce increase nuclei colonies

Second, there are concepts that beekeepers must monitor and manage, and in some ways, manage to avoid:

1.   Varroa mite population development
2.   Queen problems and replacement failure
3.   Whole hive nutrition
4.   Bee and agricultural chemicals and their interactions

In a perfect world of beekeeping, every beekeeper will maintain colonies where every production queen is bred from a diversity of varroa-mite tolerant breeder queens. Every season, these colonies are used to produce one or more increase colony (increase nucleus) for use in their operation. As beekeepers establish colonies and nuclei, they monitor each colony for threshold levels of varroa mites, signs of queen failure, signs of poor nutrition and

Adult varroa mites found on a sample of drone brood removed by a capping fork.

exposure to general agricultural and specific chemicals used inside the beehive. If any of these are detected, appropriate measures must be taken on a timely basis.

To sustain colonies, new colonies must be generated to replace the ones that are lost due to natural and human causes. That describes much of what we do in beekeeping. We keep our colony counts steady, increasing colony levels to help grow the operation, and perhaps sell surplus colonies to other beekeepers to return some of the money we put into our beekeeping operation. To develop a program of sustainable management, beekeepers must generate a great deal of good for the bees, while minimizing harm, and staying afloat financially themselves.

All beekeepers must do no harm to their bees. It is a principal of animal husbandry. There are videos out there of industrial apicultural operations where boxes are stacked up without smoke. I have been with beekeepers who kicked the lid of a hive with their foot before adding supers. No smoke was used and many bees are crushed when the supers were piled on the colony.

Beekeeper in Grenada recording observations after working new increase colonies made up to coincide with the spring and summer rains and subsequent plant development and production of nectar and pollen.

New beekeepers must learn simple things, like the proper way of using smoke to get the bees off the edges and tops of the hive bodies so they are not crushed before stacking up the boxes when finishing a hive inspection.

My personal perspective of sustainable management calls for every beekeeper's overwhelming respect for this organism, the honey bee.

Box 3. Best Management Practices and Integrated Pest Management

Many experts and world leaders call for the use of the BEST MANAGEMENT PRACTICE (BMP) in agriculture. This should apply to beekeeping as well. BMP for beekeeping is a practice, or a group of practices, shown to be effective and practical methods to improve the health and productivity of honey bee colonies while reducing any risks to them.

BMP should not be confused with IPM, Integrated Pest Management, which is a consideration of all possible pest control methods integrated together to reduce the development of pest populations while minimizing the use of pesticide amounts and types to reduce risks to human and environmental health.

IPM for beekeeping incorporates pest population monitoring, establishment of pest thresholds, learning the best time for treatment while using the safest compounds for use with honey bees. This is integrated with cultural, genetic, and mechanical practices—including the prompt isolation of removal of sick colonies, limiting the spread to other colonies and reducing the spread to reduce pest threats.

Cultural practices include drone brood removal or destruction, use of screened bottom boards, removing old combs and creating breaks in the brood cycle. Genetic practices incorporate the use of mite- and disease-tolerant queen lines to reduce pest or disease development, buildup and spread.

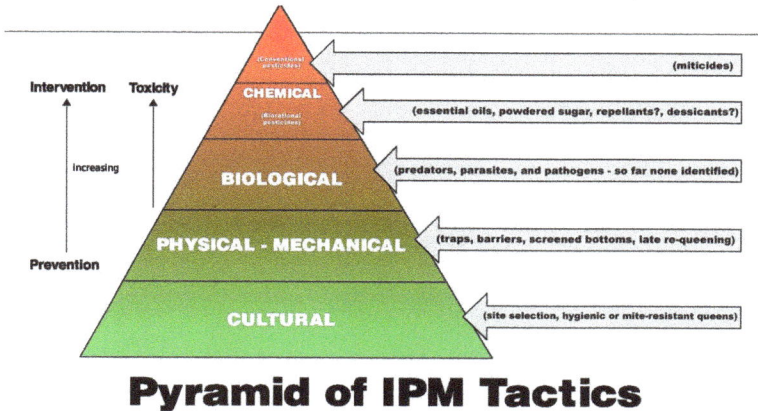

Pyramid of IPM Tactics

IPM triangular relationship of cultural, physical-mechanical, biological and chemical means of control of a pest complex. There are no effective biological controls of varroa or tracheal mites. Penn State University.

PART I
PRO-ACTIVE BEEKEEPING

The two management concepts all beekeepers need to pro-actively incorporate in their practices are to maintain and use to grow their colony numbers are the use of Mite-Tolerant Queen Stocks and the routine production of Increase Nuclei Colonies.

Mite-Tolerant Queen Stocks

The subject of using mite-tolerant queen lines was discussed in Chapter 3. There we reviewed the use of one or more of three types of methods of development of queens selected for mite-tolerance (or resistance).

The first group includes the *Survivor Stocks* (VSH, Russian and Saskatraz queen lines). The second group include those lines that were *Genetically Selected* for certain traits, such as the Minnesota Hygienic queens and the Mite Biting Bees. The third group, which is less well defined, are the lines that have *Adaptations Affecting Varroa Development*, including races with shorter worker and

119

queen brood developmental cycles (Africanized bees) and those which somehow limit mite reproduction on brood.

An essential component of keeping colonies alive is this: All colonies in an operation should contain a mite-tolerant queen laying eggs to produce workers and drones. When a queen is of a special type, say a Russian queen, something amazing happens when she produces drones—they all carry 100% of her genetic information regardless of her many drones partners. This is an advantage of the haploid sex determination of drones.

Geneticists say that these drones are like gametes (sperm or eggs) of the queen. When these mite-tolerant drones mate, they carry the mite-tolerant traits of their mother with them, and the resulting daughter workers (and queens) also carry this trait.

A mite-tolerant management plan incorporates queens from one mite-tolerant line used the first year and used to produce drones in the second year. In the second year, a group of queens from second line are installed to produce daughter queens in increase nuclei. This cycle can be repeated in the third year or a third stock introduced. This is similar to the logic used by Bud Cale with the Starline Hybrid bee (see Chapter 3), but without the inbred lines.

Increase Nucleus Production

The second essential component involves the routine production of new bee colonies. They have two advantages: to introduce a new and hopefully mite-tolerant queens, and to produce a period of broodlessness[32] that removes a key part of mite feeding—the open brood about to be sealed—and reduces varroa reproduction. While we discussed obtaining nucleus colonies in Chapter 4 as an option for starting new bee colonies, our objective here is to make our own increase nucleus as a means of stabilizing and growing our healthy apiary.

As we discussed, the increase nucleus colony has all the same elements of a full-sized colony but a fraction of the parent hive: frames of brood in all stages (eggs, larvae and pupae), frames of pollen or bee bread, frames of honey, and bees, especially young bees. To that we add a laying queen bee. Drones are not needed.

To make up an increase nucleus, a beekeeper needs to pull these components from one or more strong, healthy colonies into a new increase nucleus colony, quite often a five-frame nucleus. A standard ratio might look like this to fill a five-frame nucleus box:

- Two frames of brood, mostly sealed and emerging.
- One frame of pollen and bee bread (stored pollen).
- One frame of honey (if unavailable, use a single frame feeder with a 1:1 ratio of sugar to water).
- One drawn empty comb or a frame of foundation or a frame with a starter strip.
- One mated queen that has been laying fertilized eggs.

The first thing to consider when starting a nucleus hive is the source for a young, mated and mite-tolerant queen. The queen may come anywhere from California or Hawaii, but it must be mite-tolerant. I'd rather wait a few weeks rather than buy a queen that is not mite-tolerant. This unit can be made as early as spring depending on your source, but for me in southern Michigan, I like to start the increase process in April.

A frame of brood from a newly established increase nucleus.

Two five-frame nuclei, separated by a solid divider. Their entrances face opposite directions.

My wood five-frame nuclei colonies, each housing an instrumentally inseminated grafting mother. The entrance feeders supply fresh water in hot weather.

A second method I use to make a Doolittle Increase Colony as described in *Increase Essentials*. Brood frames are removed from one or more colonies[33] and the bees (which might include the queen) on the frames are brushed or shaken back into the hive. You do not need to find the parent colony queen. Add:

• Two or three frames of brood, mostly sealed and emerging.
• One of these frames may have one or more a queen cells, brushed so as not to be damaged, and not shaken
• One frame of honey and pollen and bee bread (pollen) from the parent hive.
• If needed, one drawn empty comb or a frame of foundation or a frame with a starter strip.
• One of these frames may have at least one sealed queen cell on the comb and positioned so it is not damaged when put into the box. If no queen cell is provided, add a caged mated queen. Lacking that, add a ripe queen cell or virgin queen.

This nucleus colony is placed above a hive with a large bee population (perhaps a strong colony with lots of drone brood and queen cups in the early stages of swarm preparations) with a queen excluder separating the colony below from the increase

I use the Doolittle Increase Method in my urban apiary. Queen-and bee-free brood frames are raised above the queen excluder over a strong hive. Bees instinctively move through the excluder to cover and feed the brood. In a few hours or the next morning, the new nucleus colony is moved to a location in the same apiary.

nucleus above. Nurse bees from below are attracted to the brood pheromone from the bee-less frames. After 4 to 24 hours, the new hive should be moved to a separate hive stand. It will not need to be moved to a new yard because all the bees are young nurse bees. Reduce the entrance because there are very few older bees ready to defend the unit. The colony will re-assign younger bees to needed tasks. Another amazing bee trait!

Once in this new location the increase nucleus should be checked as needed to ensure the queen has emerged and is laying worker eggs. Time the production of the Doolittle Increase Colony when surplus queen cells are being produced. You'll most likely find success at the peak of swarm season or while a colony is in the process of supersedure. This system works anytime of bee season.

When queen cells are used, brush the bees off the comb, do not shake them. Expect about 25% of these queens to fail to mate and establish a new hive. When this happens, combine the bees and frames back with a strong colony, which may be another nucleus where that has a queen successfully mated.

PART II
MONITORING AND
MANAGEMENT

Four vexing challenges face individual hives, local apiaries and entire beekeeping operations:

I. Varroa Mites
II. Queen Introduction and Management Problems
III. Bee Nutrition and Monoculture
IV. Bees and Agricultural Chemicals

New beekeepers should focus on becoming better beekeepers by learning how to balance varroa mite levels, mite-tolerant queens, good colony nutrition and avoiding pesticides contamination. Attempt to benefit the honey bee and the environment at all times with a side bonus: causing no harm.

Let's review each of these four issues:

Varroa Mites

Without a doubt, the biggest obstacle to sustainable beekeeping in the United States and Canada for the past three decades has been the wide-spread infestation and destruction of hives by the no good, very bad parasites known as varroa mites. The challenge is huge. Prof. Ernesto Guzman at the University of Guelph once said that 90% of the colony losses in Ontario were directly or indirectly related to varroa mites[34].

Many new beekeepers may fail to grasp the impact of varroa on colony mortality for a variety of reasons[35]:

1. Mite levels are hard to determine without intentionally sampling for them. Mites are predominantly found in two places: either inside the sealed worker or drone brood cells, or between the abdominal sclerites of worker bees while feeding on adult bee fat bodies. If you see mites on the surface of worker bees, it usually means that the colony has so many mites that it is about to

Adult worker bee with a large number of varroa mites feeding between the sternites of the abdomen. M. Creighton.

Cross-section of a varroa mite located between two abdominal sternites of an adult worker bee. The mite is feeding on the fat bodies of the bee. S. Ramsey.

collapse and die. It also means there may be no available worker or drone larval of the right age for the mites to enter to reproduce, and a majority of bees are physiologically weakened by the mites.

2. In a healthy, growing colony, mite populations may be relatively low for a majority of the season (e.g. February to September) because mite reproduction levels lag behind the production of bee brood. In the late summer and fall, however, the rate of production of new bees slows and drone brood often disappears. The resulting number of mites creates an explosion called the *Varroa Bomb*. Then BANG! the ratio of mites per 100 worker bees increases rapidly. This is when colonies start to collapse and die. It has been shown that workers from dying colonies leave and move to another colony, often carrying mites on their bodies. The weakened colonies are also an attractive source of honey, and strong colonies may send out foragers to rob them of their resources. The robber bees may become covered with mites and return to the hive. Even well-managed colonies rapidly become filled with mites from other colonies, are faced with heavy feeding and are often dead by the new year.

Key impacts including:

a. The bee colony is converting from its seasonal growth phase into a 'getting ready for winter' phase where the bee population's growth rate slows, and there are fewer worker brood cells for the mites to use for feeding and reproduction. This puts greater mite and virus pressure on the developing bees in the existing cells.

Queen and worker with adult mites on their bodies. This is not normal and is an omen for colony death. This may be associated with a Varroa Bomb where mite-infested colonies collapse, and mites migrate to healthy colonies. As a result, the newly affected colonies are usually dead by the start of the new year. File.

b. Drone brood is the preferred food for female mites and is usually abundant during the spring and summer months but is in short supply in the late summer and fall. As drone brood dwindles, more and more worker cells are infested with varroa mites. This weakens the bees, especially the bees that become the very valuable fat or winter bees, the ones that keep the colony alive during the winter. With high mite levels, the colonies often die before the start of the new year. It is a tragic loss. It happens all the time, even to my bees.

Beekeepers use screened bottom boards, sticky boards and various sampling methods to sample mite populations. Leading researchers feel that colonies should have no more than 2 mites per 100 worker bees in the hive. That means if a sampling technique shows one mite per 100 bees, the colony should be okay, but anything more than that should require treatment. Other research groups say that 3 mites per 100 is the threshold. Keep in mind that zero mites are best.

There are several rather simple methods being used effectively for mite sampling. See Box 4. The beekeeper must decide if she

Varroa mites are attracted to drone brood (but older than what this photo. shows). Mites are attracted only to larvae that are close to the cell-sealing stage. They hide inside the drone cells in the brood food until the workers seal the brood. Once the cells are sealed, the mites feed on the drone pupae after their final larval metamorphosis.

The mites leave the drone brood cells when drones emerge. A. Connor.

One popular method to control varroa population is to remove drone brood after the cells are sealed and the mites are trapped inside the cells. Some beekeepers remove drone brood and freeze it to kill the brood and the mites. They then return the thawed frame to the colony where the bees pull out the dead drones. Here, we see a medium brood frame placed into a deep hive so the bees built drone comb in the space. This drone brood may be cut off and fed to the chickens or buried.

should use soapy water or powdered sugar, but once a process is selected, stay with it. The method should not be changed or you lose the ability to make accurate comparisons. Take samples of young worker nurse bees, the most attractive to the mites, from brood frames and allow any older bees to fly out.

Extensive studies have been made to use Integrated Pest Management techniques to control varroa mites. Unfortunately, we still do not have an effective biological control agent to control varroa mite populations, although researchers are actively attempting to develop one. We are limited to mite trapping (often using drone brood), screened bottom boards and sticky boards, making increase colonies to create a break in the brood cycle, and using mite-tolerant queen stocks to reduce mite survivorship.

Most commonly, beekeepers control mites by using chemical treatments in bee colonies. These are summarized in Box 7. Please check with the Honey Bee Health Coalition website (honeybeehealthcoalition.org) for the latest information.

As stated above, the best way to control varroa mites is to integrate mite-tolerant stock with the production of increase colonies. While I leave screened bottom boards on some colonies all year long, I no longer use them for mite sampling. In a perfect

By combining screened bottom boards with powdered sugar dusting, many beekeepers reduced mite populations. This system works best when repeated several times during each 21-day worker brood cycle to 'catch' the mites.

With a screen or queen excluder over the combs, powdered sugar is spread over the tops of the frames, using the screen as a means of spreading the powdered sugar.

world, I use the powdered sugar shake method every 4 to 6 weeks throughout the growing season to see if mite levels are too high. Many people use the new Mite-Checker tool and find it essential to take the kit to their apiary.

The strength of this system is the use of mite-tolerant queens (open mated queens grafted from queens selected for mite tolerance). In the past ten years, I have used Russian, VSH, Minnesota Hygienic and Saskatraz queen stock successfully. The Purdue mite groomers have not become established in my colonies, which is something I hope to correct. The weakness of this system

Screened bottom boards are used to monitor mites as part of an IPM program. The colony on the left has the sampling tray removed for analysis.

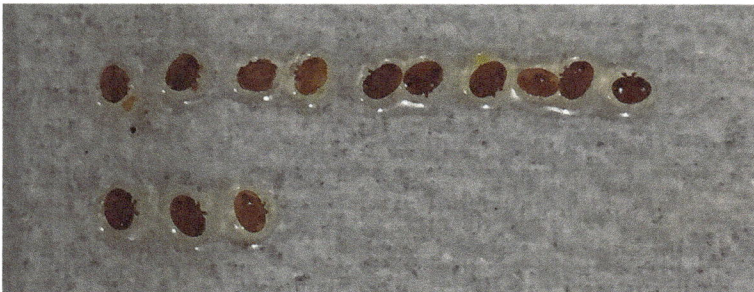

Varroa mites on a microscope slide from the Purdue University mite-grooming (MBB) study. They will be examined to see if their legs and antennae were chewed by worker bees. These mites are from natural mite-fall (no treatment).

129

Box 4. Sampling Using Powdered Sugar

UL: Shake bees from the brood nest into a container. Try not to include the queen. UR: Measure out 1/2 cup of bees into a jar. ML: Add 2-3 tablespoons of very dry powdered sugar (no lumps). MR: Shake bees vigorously until all mites detach from the bees. LL: Let bees sit for several minutes, then shake some more. Afterward, shake the sugar and mites onto white surface. LR: Mist with water to dissolve the sugar, leaving the dark mites visible and easy to count.

is that my travel schedule interferes with routine sampling. I am sure more colonies fail to thrive because of this and my refusal to treat. I am also behind on culling out old comb, which should be done.

Queen Problems and Replacement Failures

I have put the combined issue of queen problems and replacement failures into my number two position of the many challenges beekeepers face when dealing with queens. While many beekeepers feel that the primary reasons their colonies have died is because of either varroa mites or starvation, they often overlook a variety of issues associated with queen failures. Many queen failures occur during the summer and fall when hives are stacked with supers and it is difficult to check the brood nest. If the old queen is superseded and the daughter queen fails to mate—none of this is seen by the beekeeper. The key issue originated with a queen's mating failure may not be seen until next spring. This is often the case with drone-laying queens and the pitiful colony remains.

Double jar system for shaking mites off bees. Similar to the powdered sugar system but more accurate. Hardware screen separates the two jars. Soapy water, antifreeze or washer fluids are used as the liquid. Disadvantage: Bees are killed.

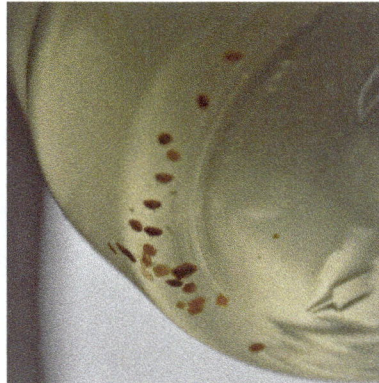

Once the double jar has been shaken following the suggested interval, the varroa mites dislodge from the bodies of the worker bees and can be found in the bottom of the bee-free container. They may then be counted.

Mite-Checker tool used in mite sampling. It does the same thing as the jar with a hardware cloth lid shown in Box 4.

Every year the National Honey Bee Survey (conducted by the Bee Informed Partnership and USDA Animal Plant Health Inspection Service) examines colony health throughout the United States. It began in 2009 to figure out why colonies were dying in the country. It also surveys for potential threats and colony pesticide analysis. They have observed the extensive and pervasive nature of summer colony losses and write: "Over the last 8 years, winter losses have been unsustainably high ranging from 22% to 36% nationally. In the previous 2 years of the BIP Colony Loss survey (2014-2015 and 2015-2016), *summer losses equaled or exceeded winter losses*." (Emphasis mine)

Summer losses are related to a wide range of challenges, including but not limited to pesticide exposure, starvation due to monoculture, queen mating failure, the resulting drone laying workers, and queen introduction issues. Of these, let's look at the most pressing queen issues and how they impact colony survivorship. The first issue is involved in work that indicates that queens have trouble during and after their introduction into the hive.

Introduction and Establishment Failures

Early on in my beekeeping career, I learned that queen problems have been an industry concern for a long time. In one of my many conversations with honey bee geneticist and USDA bee breeder Otto Mackinsen, while he worked at the Baton Rouge Bee Laboratory, he described a study he worked on during the 1930s, when queens were raised in the USDA Bee Lab in Louisiana and used to establish package bees. The packages were assembled and shipped to the USDA Bee Lab in Wisconsin (now closed). These were used to establish production colonies there. Their

Grafting tool removing a larva from a brood cell. The larva is placed into a queen cup and put into in a queenless colony where the bees feed it royal jelly, resulting in a queen.

performance was monitored carefully. At the end of the summer, the researchers reported that about 35% of the original queens were gone, either rejected during package installation or replaced (superseded) during the summer.

Since then, some bee researchers have understood that queen bees are not accepted at a high level, not at all times and in all years. The USDA study was done before tracheal or varroa mites had entered North America. There are years when queen and drones are produced under ideal beekeeping conditions, and they mate well. In other seasons, there may be cold and rainy weather patterns that interfere with colony foraging, queen and drone production, and ability to mate successfully. Fortunately, even a few hours of flight time in the afternoon may provide adequate mating conditions.

There are undoubtedly many factors involved in this. Let's look at two factors—queen weight and queen race—where data exist concerning queen acceptance, performance, and failure.

Low Queen Weight and Thorax Width

A 2015 Egyptian study [36] examined weighted newly emerged adult queens and then introduced into mating colonies. The resulting differences were huge based on emergence weight, dramatically influenced mating success.

The successful mating rate of larger queens out-numbered the small and medium queens. Larger queens mated with a success rate of 79%, 14% failed to mate and 7% became drone layers (unmated but alive with the queens producing unfertilized eggs). Small queens mated with a success rate of only 29%, had 22% drone layers and 49% mating failure.

My conclusion: small queens failed to mate three times more often than large queens, By doing so, the small queens failed to produce diploid daughters, suggesting that small queens have unknown mating issues. Added to that is the fact that small queens produced 22% laying workers, a huge cost in new colony development.

The good news is that in this study the bees produced more large queens than smaller queens. But are small queens unable to complete during the mating flight? What factors determine their larger failure rate?

Beekeepers should use only large virgins for mating units. The virgins in this study were weighed within 15 minutes of their emergence from their cells, a practice most queen producers would find difficult to follow.[37]

But we may have a solution, based on another study[38] made in North Carolina that examined many queen characteristics, including queen 'wet weight', the width of their thorax and head width using a digital caliper. Interestingly, queens from the Southeast were significantly larger than those from the West. The averages were:

Wet weight of non-laying queens 184.8 ± 21.67 mg
Thorax width 4.35 ± 0.188 mm
Head width 3.62 ± 0.123 mm

Table 3. Queen Weight and Mating Success

Queen Weight	Number of Queens	Successful Matings	Drone Layers (%)	Failures (%)
110-130 mg	45	13 (29%)	10 (22%)	22 (49%)
140-160 mg	68	31 (46%)	11 (16%)	26 (38%)
Over 160 mg	130	103 (79%)	9 (7%)	18 (14%

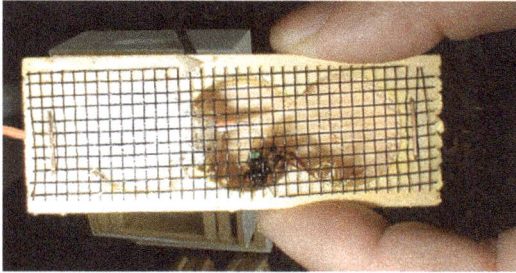

A dead queen in a shipping cage. When this happens, immediately contact the supplier and arrange for a replacement queen. S. Repasky.

These are potentially important findings. Once perfected, measurements of thorax width could lead to the development a standardized measuring systems used when pulling queens from mating nuclei. This is usually the first and only time the queens are handled, so the use of a small digital caliper used in crafts and jewelry making makes sense. Unlike queen weight, the width of the thorax does not change during a queen's life

Experiences Using Small Queens

While working on the mass instrumental insemination of queens in south Florida, I had the opportunity to handle thousands of these royal ladies. We used several methods to ensure our queens were as large as possible. First, we used the smallest and youngest larvae during grafting. We timed grafting 3.5 days after the breeder queen was introduced to the comb. One employee worked 7 days a week just to move empty comb consistently at the same time every day; we wanted to have larvae of the right age—12 hours post emergence—to produce optimally sized queens.

Second, everything from the breeder colony where this queen originated to all the cell production and holding colonies, were fed 1 to 1 sugar syrup, even during the sporadic nectar flows south Florida experiences. Protein feed was given early in the season.

Third, we learned to use our fingertips to detect each queen's size based on the thickness of her thorax. We dispatched the small ones into a jar of alcohol, keeping them out of production. We became a bit fanatical about queen size and it paid off when our

Judging the size of a queen by the size of the natural cell (L) or raised cell (R) is common, but it is not a guaranteed way to predict overall queen size and performance. Weighing queens at emergence or measuring the width of the thorax is more accurate.

queens were bigger and easier to handle during insemination. Small queens were a real challenge to inseminate.

Some of our inbred lines routinely produced smaller queens than the other lines, suggesting a genetic link to queen size. They were productive, and I felt they did not last as long and required more expense to maintain them, but I recorded no date on this.

Work of David Miksa in Florida on Cell Size

I wrote about Dave and Linda Miksa in Loveland, Florida in my book *Bee Sex Essentials*. Dave and Linda obsess about queen size too. The Miksa family produces and sell thousands of queen cells every year, and focuses on producing big cells. Dave, too, emphasizes using the smallest larvae he could during the grafting process. To ensure large cells, he has the cells and cell bars dipped into hot beeswax until coated up to the lip of the cell. No wax was allowed to enter the cells. By supplying abundant beeswax to the base of the queen cells, bees were stimulated to produce larger

Plastic queen cage with cap removed. This allows the worker bees to remove the sugar plug and liberate the queen. This process takes 4 to 24 hours.

cells. To ensure the developing queens inside these cells were large, Dave feeds each production colony a rich mixture of sugar, protein and a variety of supplements, producing large queens. Queen quality is easily improved by using only large, well-mated queens. These queens will produce larger colonies and, in turn, be more productive pollinators and nectar gatherers.

Impact of Racial Origin on Queen Attractiveness to Worker

What happens when you try to introduce a new queen into a colony filled with bees from another race? Beekeepers often say they have problems introducing new and unrelated queens into a colony, and the same Egyptian study[39a] cited above shows that. Related queens experience greater acceptance in hives populated with bees of the same race. Three races were used: Egyptian, Carniolan, and Italian races. The researchers found that when related queens were introduced to the worker bees in a hive, acceptance was better, compared to when unrelated queens were introduced. We know that queens produce an odor (queen odor or pheromone) that has a genetic basis. Worker bees detect these differences when a virgin queen is introduced into a hive. The Egyptian researchers found that unrelated queens receive more aggressive contacts when compared to sister queens, suggesting that the queens were producing pheromones that triggered a negative response by the workers. Or did they lack a component that inhibited attack?

With African honey bees, other researchers have shown that queen acceptance was highest in the queens that emerged sooner, piped more, killed sister queens and received more vibrations from workers.

My Florida Observations Introducing Unrelated Queens

In my Florida work, I inherited an alphabet soup of inbred lines that taught me a lot about the role of genetics and queen introduction. My biggest takeaway was that unrelated queens were much more difficult to introduce to colonies than related queens.

The Starline Hybrid was made from four lines: H, G, aE and F, making the cross HG x aEF or aEF x HG. The F line was a critical line, but it was the most distantly related to three lines that had Italian roots. The F line's origin was from a single Carniolan queen found in California in the 1940s. Bud Cale outcrossed the queen's daughters to good Italian stock and then backcrossed the line to the characteristics he desired, selecting the yellow queen color commercial beekeepers demand in their queens.

Large queen cells from Miksa operation in Florida. Many efforts are taken to ensure large cells.

Different queen color patterns may reflect vastly different genetic backgrounds. When trying to introduce bees of one race, it is best to use bees of the same race or you may experience lower introduction success.

This F line was a critical part of the production of the hybrid, and we needed between 50 and 100 colonies with pure F queens to produce drones. These queens carried their Carniolan traits and, I'm convinced, their unique Carniolan pheromone signature.

Whenever we attempted to introduce F queens into non-F line colonies, we delaying introduction by holding queens in their cages for a more days, feeding the colonies thin sugar syrup and reducing the amount of smoke we used. It was a challenge.

The bees from the F line were wonderful to work with. We considered them productive and gentle, and the queens were long and graceful. There was never discussed replacing the line because we had nothing that equalled the production and gentleness of these ladies. The colonies were very sensitive to environmental conditions and would shut down egg laying when the nectar flow was over while other inbred lines charged ahead as if the nectar flow had not stopped. This was an essential part of the line's contribution to the hybrid, serving as a genetic throttle on the powerful engine of the hybrid bee.

Ultimately, we always tried to have bees in new colonies receiving F lines to carry some of the F line genetics in the worker bee's genetic pedigree. If we put F line queens into Starline colonies, we had good acceptance, so we learned to keep our own hybrid colonies in production at all times just to support this one inbred line.

Ideal Introduction Method

Vermont beekeeper and queen producer Michael Palmer uses a queen introduction system similar to mine. He uses push-in cages, about 4 x 6 inches in size. He places the cages above emerging brood and cells filled with nectar. He installs the new queen during the same visit that he removes a queen, either for sale or use elsewhere. He puts the queen under the cage without workers, gently pushes the cage into the comb, and leaves it there for four days. Before removing the cage, Palmer checks for eggs outside the cage to make sure that a second queen is not present in the hive. He checks to see the queen is on the comb rather than on the cage where she could fly away.

The system allows direct bee-to-queen contact for maximum pheromone transfer, proper feeding of the queen to initiate egg laying, and provides the opportunity for the queen to lay eggs if

One inbred line was difficult to introduce into other bee stocks. It was used in the Starline hybrid during my tenure running the bee breeding program established by Dr. G.H. Cale Jr.

Queen without introduction issues.

there is room for her to do so. These methods help introduce a new (and often expensive) queen into a colony where a queen is already present.

Queen Nutrition

During larval development and as an adult, queen feeding is essential for full development and maximum egg laying. In the past, feeding was something beekeepers did only in the fall to replace honey that was harvested, and in the late winter and spring to build colonies in number so they could be used for queen production or for making increase colonies or packages. We recognize the need for extra stimulation for good queen cell production—the larvae must be well fed and the young queens must receive optimal nutrition during the period from emergence to mating. High nutrition is needed to ensure a high egg-laying rate and brood nest expansion. Both sugar syrup and protein patties are recommended for good queen nutrition.

Queen Status

Queen bees change their pheromone 'signature' as they pass through different stages of life. The process starts slowly with queen cells.

Queen cells about two days from emergence attract worker bees (receptor bees) that often chew the wax off at the tip of the cell, leaving the queen pupa's silk exposed. After the queen emerges, her queen pheromone production increases as she ages. One research paper showed that about 12 hours after they emerge, queens have a surge of pheromone production, but pheromone production continues to mature for a year or longer.

Plastic queen introduction cage.File.

141

Mate Number

Perhaps the most critical time of a queen's life is the response she receives from the bees immediately after mating. Dr. Elina Niño, now at the University California, Davis, showed that the chemicals in the queen pheromone changes based on her physiological state as influenced by the number of drones she had mated with. "Workers exhibited the strongest attraction to the pheromones from highly inseminated queens, compared with those injected with less semen or with plain saline or with those that were virgins," Niño said. "She's also telling the workers that she's a mated queen, and that she's either poorly or well mated."

The queen pheromone has several roles, including to form a retinue around a queen, a group of worker bees that function to move the queen pheromone throughout the colony. The pheromone increases foraging, inhibits queen rearing and inhibits worker bee egg production (which results in laying workers if the queen is absent).

Mating Failure and the Role of Laying Workers

Through her pheromone production, a good queen suppresses egg laying by her sister and daughter worker bees. This seems like a simple statement, but we need to understand what it means.

Pollen contributes directly to royal jelly production, reflected in the level shown in these queen cups.

First, every worker bee has the potential to produce eggs but only a few eggs per day. This behavior usually only lasts a few days or a week at most. But hundreds, if not thousands, of worker bees can produce eggs, and a colony with laying workers will have thousands of worker eggs. These worker-laid eggs do not seem to be recognized by other worker bees, so multiple worker eggs are often deposited into the same brood cells. Eventually, one will be selected, and the rest are removed. When only one egg is left in the cell, worker bees will form a bullet-shaped capping that will accommodate the larger size of the drone.

Because a worker bees never mate, her eggs are haploid and turn into drones. When produced in a worker cell is will be diminutive in size. These worker-sized drones produce viable sperm.

None of this would have happened if the queen were present and producing queen pheromone. Both the queen pheromone and the brood she produces suppresses worker bee egg production. But the production of unfertilized eggs by workers is rare, right?

Not really. When I had the pleasure of working with Nikolaus and Gudrun Koeniger on *Mating Biology of the Honey Bee*,[40] they discussed the percentage of mating success in queen colonies. In the work they have done, and in research facilities in Europe, they found that 75% of all queens put into mating colonies were successful in mating and laid worker eggs.

This means that 25% of all queens failed! Why? Because of weather and the intervention of predators like dragonflies and king birds. That means that 25% of all mating colonies, full-sized colonies undergoing supersedure, recently established swarms replacing the old queen—all of these colony units are subject to the same 25% 'rule'.

Unless a beekeeper happens along at the right time and has the right information, these colonies will become hopelessly queenless and then broodless and then have laying workers.

A laying worker colony may appear to have many worker bees, but they are extremely difficult to save. Combine with a strong queenright colony and let them sort out the pheromone issue.

When the queen was not accepted in a new package colony, laying workers eventually developed and filled combs with drone brood. J. Rigney.

It is my observation that many colony losses experienced during the growing season are a result of this high level of queen loss and is part of normal queen supersedure. It is risky for the bees. Without human intervention these colonies will die, perhaps during the winter, with just drone brood appearing among the dead bees. Without a mated queen there will be no new young worker bees.

Combining low mating success rates, small queen problems and rejection of unrelated queen stocks, it is no wonder that summer and early fall losses are a vastly under-recognized issue in keeping colonies alive.

PART III
WHOLE HIVE NUTRITION
DURING BEE DEVELOPMENT

Bee Colony Nutrition

Recognition of the role of honey bee nutrition has grown in importance as an issue facing beekeeping. In the past, most beekeepers did not consider taking an active role in feeding their colonies because natural bee-collected food seemed abundant. I remember instructing beekeepers not to feed in the late winter or else the bees would grow so strong they would swarm. Yet,

The addition of a frame of brood with bees is often the solution to hive challenges, such as a failing queen, low bee populations (common in package hives), laying workers (sometimes), and other issues. Always check for the queen!

Box 5. Functions of Support Nuclei Hives[41]

About three months after your initial colonies are established you can set up a nucleus colony, assuming the season's nectar flow and colony development allow. Learn from this smaller colony and evaluate how well it survives the winter. If all goes well, you can either keep it or sell it; overwintered, locally-produced colonies demand a higher price than nucleus colonies trucked in from the south.

If you produce smaller colonies, you will sustain your home-based beekeeping operation by effectively reducing the number of varroa mites in a colony by creating a pause in the egg-laying of the queen, often called a *break in the brood cycle*. During this time the varroa mites stop have fewer brood cells to infest and the bees may use grooming behavior to eliminate the parasites. Having a few of these increase colonies provides a backup queen or even a backup hive, just in case something happens to the original hives. Finally, having a few increase colonies may provide you with additional income to support your beekeeping interests.

Box 6. Photos of Queen Management

Worker feeding queen larvae in this apparent supersedure cell on a frame of worker brood. U. Calif., Davis.

Worker bee adults with deformed and k-wings, caused by viruses, likely transmitted by varroa.

Decomposing queen larva killed by black queen cell virus. This virus is present in low levels in all colonies.

Naturally mated queen with a reduced egg-laying rate was observed expelling multiple eggs after a worker bee vigorously whacked the tip of the queen's abdomen with the worker's hind legs.

Recognition of a good colony.

Sealed and emerging worker and drone brood.

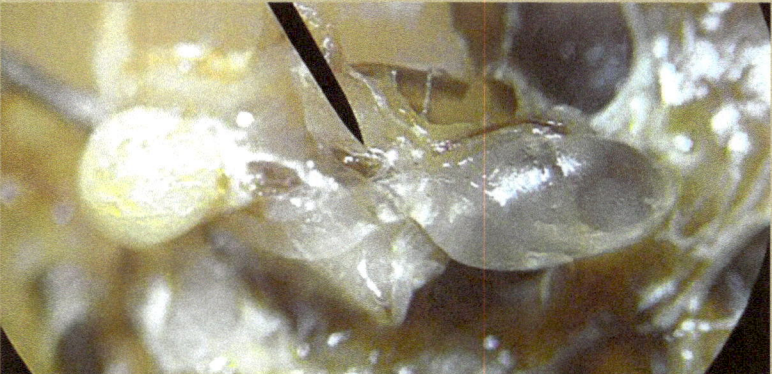

Dissected queen with spermatheca on left, covered with a network of tracheal tubes that provide oxygen and remove waste gases. The clear venom sac appears on the right.

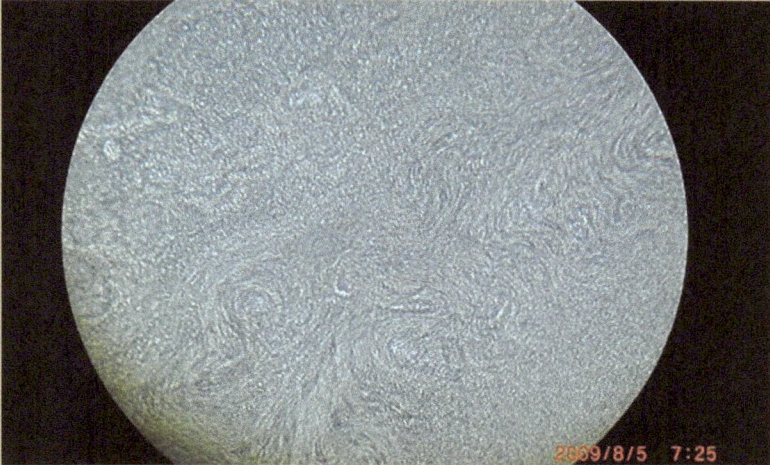

Sperm supplies start at 3 to 8 million in a newly mated queen and fall continuously as a queen lays eggs. Queens appear to release a volume of spermathecal contents, not a set number of sperm. As a queen ages, the appearance of drones in worker cells is a good indicator that the sperm concentration has fallen. L. Houston.

Two queens, one yellow, one dark, on a frame of emerging brood, including drones. About 15-20% of spring colonies have two queens, often mother-daughter combinations. This is common in supersedure. File.

Stay calm and be systematic when searching for a queen.

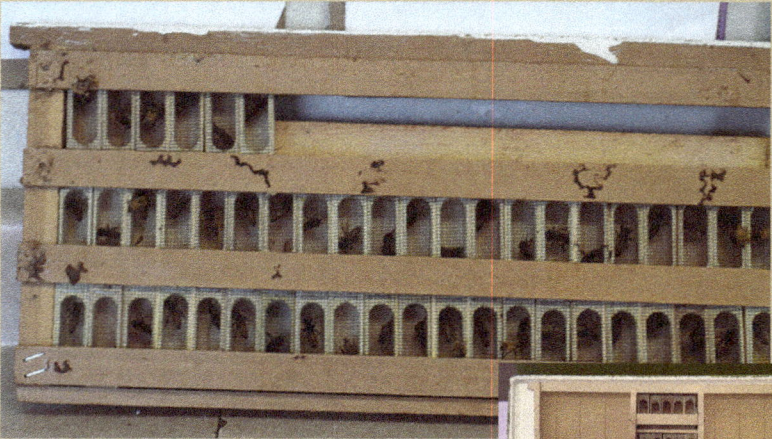

L: Queen bank frame holding mated queens.
R: Holding frame suitable for keeping smaller numbers of queens in a queenless nucleus hive.

Drone larvae on special green drone foundation.

Brood of different ages. Some would consider this spotty brood.

Supersedure cells along with reduced brood nest and backfilling with nectar, a typical sign the queen is being replaced.

Supersedure cells on deformed comb.

New cell, poorly mottled with wax. While some beekeepers avoid this lack of wax, it may just be an aspect of a strong nectar flow.

Frame of young larvae after the colony's queen was removed, resulting in the start of multiple queen cells. This is the emergency response. D. Caron.

A queen cell produced by a beekeeper who then installed it properly on a frame. Check back in 3 to 5 days for the queen's proper emergence. A hole in the cell's side means another queen was already present and killed the queen before she emerged.

Queen cell with cut mark where a queen has tried to emerge. Workers added wax externally to keep her in the cell. Bees do this to coordinate the release of the queen to coincide with production of an adequate number of bees for swarming. Workers feed the queen through the opening.

153

48-hour queen cells may be shipped.

The high cost of mated queens deters beekeepers who cannot afford queens with mite tolerant genes. An alternative to a mated queen might be a 48-hour old queen cell, based on the time of grafting. These queens are quick to produce and put the burden of cell-finishing and new-queen mating to the user of the cells. At 48 hours a queen cell is nearly filled with royal jelly, but the larva is too small to wiggle out. At this stage, the cell may be carried to another location without bees and without temperature regulation outside of 'room temperature.'

My first exposure to these cells was during a visit with Dr. John Kefuss in Toulouse, France. He packages the cells without bees in Styrofoam strips, bundles them in narrow bread wrappers, and then secures them in a box. He meets the 11 pm high-speed express train, and by 7 am the next morning, the package is in Paris. There, beekeepers install them into mating nuclei.

It takes a minimum of 350 bees to make a queen cell, and the nuclei the beekeepers were using had over 1,000 bees. This system works. I have used this system with five-frame mating nuclei with at least 3,000 bees inside and usually many more. When queenless and filled with young nurse bees, they build nice cells. John sells these cells at a much-reduced cost, since he had 'only' invested 48 hours rather than the time and energy required to produce a mated queen. The recipient beekeepers were receiving some of John's high-priced varroa-mite tolerant stock at a great price break.

I shared this method with a number of beekeepers at training classes. One of them, Dwight Wells of western Ohio, carried six of the cells from the family farm in southwestern Michigan to Columbus, Ohio and produced five queens. He was hooked. Now, he and Dorothey Morgan, president of the Kentucky Bee Breeders Association, use this method to share queen lines that carry the mite-biting behavior seen in the Purdue stock and queen families collected from the wild.

If, like me, you are an urban beekeeper with a few hives in the back yard and one breeder queen in a well-managed nucleus, you can become a source for queen cells for other beekeepers. If you spent $500 for one of John and Carol Harbo's VSH breeder queens and produce 50 cells in one graft, you could arrange a weekend pickup by local beekeepers. At $5 to $10 per cell, you could recover the cost of the breeder queen in one or two grafts! More important, to me anyway, is that you have put 50 or more mite-tolerant queens into beekeeper's apiaries, where everyone benefits.

while I was at The Ohio State University, geneticist Dr. Walter Rothenbuhler took on the task of researching the 1970s problem of Disappearing Disease. After some research, he determined that this was not a genetic issue, but a nutritional one.

Assuming that nature provides all the nutrition that hive bees need is probably still the case for many beekeepers. In my small backyard setup, the bees get to keep all the honey they made during the season on the hive for the entire winter. Surplus honey in the spring may be harvested or used to establish new colonies. The colonies have to be checked for food reserves and mite populations, of course, and assessed for adequate reserve stores.

In my urban Michigan location, I have observed pretty good pollen supplies at all times of the bee season. If I had my colonies

A green bee on goldenrod.

Sunflower

Almonds are planted in rows of different cultivars to provide compatible pollen for cross pollination. These bees were placed on the pollen row, but the other cultivar has finished blooming!

155

a few miles away in the southern part of Kalamazoo County, I am sure the situation would be reversed. This is an area where there is extensive seed corn production. While there are small forests and some streams that run through the region, the dominant land use is for corn seed production. When corn is in bloom it produces a great deal of pollen, but the rest of the year the amount of pollen is limited because of herbicide applications to these GMO plants. Of course, growers are going to apply insecticides whenever there is a risk of an insect pest outbreak because of the value of the seed.

If I lived in this area, I would not keep bees there. The threat of starvation is real. While bees can fly three or four miles to get to non-corn areas, it is a region with too many risks with very few benefits to balance the scale for the bees.

Earlier, I wrote that I have used winter patties, which are mainly sugar and a smaller proportion of protein. These continue to work for me here in town. But in the southern part of the county I would feel forced to use both sugar and protein patties starting the summer to ensure my colonies are alive in the next spring. Plus, the chance of a bee loss due to insecticide use would be constant.

Monoculture and Lack of Floral Diversity

The corn zone in my county is an example of a human-made monoculture. In the California almonds, those trees create another human-created monoculture. In many cases a monoculture implies

Purple loosestrife.

Yellow sweet clover.

Dandelion.

Spotted knapweed.

that the bees must be moved in and out to another location for the rest of the season.

Agriculture is shifting more and more to monoculture for corn, beans, cotton, sugar cane, oranges and other crops. Success breeds mimicry. As one farm is successful with a crop in monoculture, other growers will sow the same seed and use the same agricultural methods. Whatever the plant is, where one farm does very well others soon appear, and the diversity of the food supply for the bee colony drops dramatically.

Beekeepers often like monocultures for their bees IF the plant is a good nectar source. Monocultures of sweet clover, alfalfa, citrus, blueberry, canola, sunflower, and other selected plants are the basis of the largest honey crops made in the United States and Canada. Beekeepers are willing to move bees for a nectar-rich monoculture.

Feeding Methods

Feeding is an important part of growing new colonies—packages, nucleus hives and swarms. For many small-scale, backyard beekeepers, the easiest feeder is a frame of sealed honey from the comb storage or from a strong hive in the apiary. It is always amazing to me to see the benefit of extra food on a colony. For full-sized colonies, there should always be a minimum of three frames of honey in the hive. Always.

In nucleus hives with five to ten frames, there should be one or two frames with sealed honey and pollen at all times.

For a rapid boost to a special colony, one I may want to produce queens and drones from, I will place an extra box of honey or

157

selected frames UNDER the broodnest. The bees respond to this by moving the honey from below the queen's activity either into it or above the brood area, where the colony will most likely expand. This system is great if you have frames with granulated honey that cannot be extracted or sold as comb honey.

Planting Trees and Forbs for Bees

Many beekeepers plant trees and forbs for their bees. From a city location to a large ranch, beekeepers have taken on the task of identifying tree and forb species that benefit honey bees and other pollinators. Some have benefited from the 'Save the Monarch Butterfly' movement since milkweeds are excellent food sources for bees.

More people should consider planting a hedgerow of trees for bees or line their suburban street with nectar producers, or fill city parks with bee friendly trees and shrubs. Every suburban gardener may put in a few basswood trees (*Tilia*), and plant some low-growing fragrant sumac (*Rhus aromatic*) as a weed barrier and ground cover around these nectar producing trees. This short sumac grows only 1.5 to 2 feet tall and attracts bees, butterflies and birds. All sumacs are clonal, and spread underground, creating a four to six foot expanse of dense foliage, flowers and later, sumac berries. Remove the fescue in your yard and put in these plants that feed the bees in the spring and turn red leaves in the fall.

Warn people not to rely upon the list of showy flowers for pollinators they find at the garden centers, from the extension service, or online, unless the list has been reviewed by someone who actually knows something about honey bees and the flowers they visit. Usually, these lists focus on butterflies, not honey bees. Others are not all that attractive to honey bees. For example, the flowering dogwood (*Cornus florida*) is a wonderful understory tree, but it produces no nectar and is only a very minor pollen source. An excellent substitution would be to plant redbud, as there are both native eastern and western species. An eastern redbud (*Cercis canadensis*) grows over my driveway.

The western or California redbud (*Cercis orbiculata or Cercis occidentalis*) is found across the American Southwest, it is cultivated as an ornamental plant and tree, planted in parks and gardens, and as a street tree. It is drought tolerant, native, and survives periodic burning.

Consider planting red maple, basswood, tulip popular, black locust, sumac, mesquite, fruit trees, willow, eucalyptus, rabbit brush and other tree and shrub species as an investment in our planet, and as a way to help support hungry honey bee colonies.

PART IV
BEE AND AGRICULTURAL CHEMICAL INTERACTIONS

Sustainable bee management avoids or reduces chemical treatment of the hives, but not so much that one is obsessed and refuses to treat bees and watches the bees die. Choices of chemical compounds permit beekeepers to use naturally occurring, albeit highly toxic molecules, to control hive pests, especially varroa mites. Development of compounds that duplicate the stimulus of brood and drone brood pheromones will lead to a new generation of mite control chemicals—molecules so specific that they interact only with the behavior of the bee pest and nothing else.

Let's look at how agriculture and bee-related chemicals impact our honey bees. All these chemicals are grouped into the class of pesticides, or chemicals that kill pests. These are divided into other groups based on the target organisms they affect:

Herbicides—The Plant Killers

Chemicals used on agricultural fields and in urban lawns are designed to kill plants. Some are designed to kill broad leaf plants, so we can grow grass in our lawns but not dandelions. Fescue and other grasses are the most commonly managed crop in America: the manicured lawn.

In the past, corn and bean fields would be filled with weed plants, especially some of the major summer pollen and nectar sources like smartweed, knotweed, thistle and milkweed. Agricultural herbicides are now used to selectively kill weeds but allow crop plants that carry a gene or gene complex to keeps it alive. These are called Genetically Modified Organisms (GMOs). Now, fields are sprayed with herbicides that do not kill the genetically modified crops. Honey bees, native bees, monarch butterflies and other animals suffer from the GMOs.

Europeans have widespread bans on GMO plants, but US government policy pretty much protects the chemical companies and the agricultural industry, but not the consumers. Bees are hurt by the loss of forage when weed plants that produce food for pollinators are killed. Since they are beneficial to pollinators, we should stop calling these forbs weeds.

Fungicides—Slayers of Parasitic Fungi

Fungicides are chemicals that kill fungi. Many common plant diseases are caused by fungi. There are implications that fungicides,

Herbicides like Roundup have become part of the GMO monoculture, where plant diversity is limited to crop fields with the herbicide killing all but the genetically modified plant species. File.

Box 7. Chemical Control Compounds

(Honey Bee Health Coalition[42])

Varroa mite chemical control can be divided into two groups, the synthetic molecules and organic molecules that occur in nature although that does not mean they are not without risk to bees or beekeepers. Most are not produced from natural materials but are the result of commercial chemical manufacturing.

Synthetic Miticides

- Apivar® (amitraz)
- Apistan® (fluvalinate)
- CheckMite+® (coumaphos)fn.

Organic Miticides

- Essential Oils
- Apiguard® or Thymovar® (Canada) (thymol)
- Api-Life Var® (thymol + eucalyptol, menthol, and camphor)
- Acids
- Mite-Away Quick Strips® [MAQS®] of Formic Pro® (formic acid)
- Formic Acid 65%
- Oxalic Acid
- HopGuard® II (hops beta acids)

Cultural and Mechanical Practices

Methods recommended for varroa mite suppression:
- Break in brood rearing, making increase nuclei colonies
- Drone brood removal
- Brood interruption (splits, nucs, indoor overwintering)
- Sanitation (comb culling/biosecurity)
- Screen bottom board
- Heat (106°F (41°C) for four hours) or MiteZapper®

which carry no warning label to protect bees, work to impact bee development or behavior. They may have a synergistic effect when combined with other chemicals.[43]

Insecticides—Chemicals that Kill Insects

If you use a solution of dish soap in your kitchen to spray on your vegetable garden to control aphids or other pests, you are using it as an insecticide. We live in a world of insecticides in most of agriculture. The organic food movement is becoming very strong as consumers are willing to accept slightly less than perfect food to minimize chemical exposure.

Miticides—Pesticides that Target Mites

American beekeepers have two species of mites: tracheal and varroa. Each has targeted chemicals. Chemicals used by beekeepers to control these two mite species can cause damage to bees, brood, queens and drones. The drone damage includes sperm mortality and reduced mobility. Queen damage includes reduced life-span (early supersedure) and lower egg-laying rates.

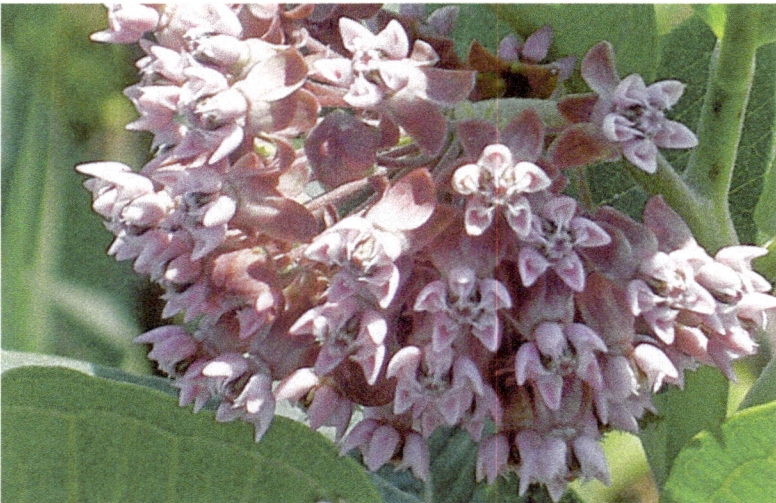

Milkweed commonly grows in agricultural crops, but its elimination by Roundup and other herbicides has reduced nectar production for honey bees and monarch butterflies.

Varroa Mite Control

These ectoparasites are so successful in damaging honey bees that most beekeepers have extensive trouble combating them. Varroa mites are discussed in several sections of this book.

Tracheal Mite Control

The honey bee tracheal mite, *Acarapis woodi* (Rennie), was first found in the United States in 1984. There were widespread colony losses throughout the country. The mite lives in the thoraxic trachea (breathing tubes) of the bee. Mated female mites crawl into the trachea through the spiracles of the thorax. In heavy infestations they show up as darkened trachea while uninfested tracheae are clear and white when viewed under the microscope. Interestingly, as rapidly as it appeared as a major problem in the United States, it became last year's problem. Acaricides (pesticides that kill mites) are available. Menthol is often used in the United States. Formic acid has been approved in some countries for tracheal mite control.

Fungus infection on a plant bug. Fungi may be considered both beneficial and detrimental depending on where and what they invade. Various insects and mites are killed by fungi. Why can't we find one that kills varroa mites? Univ. California.

PART V
WHY KEEP BEES?

For thousands of years, humans have put bees into containers and attempted to manage and manipulate them. Humans keep bees to obtain honey, wax, propolis, brood as food and pollination. We have known about the importance of bees as pollinators for several hundred years. But we have been making mead from honey for a long time—for at least 5,000 years—and perhaps it was the first alcoholic beverage, which had spiritual and magical properties.

Do bees benefit from human interaction? In their own world bees face many evolutionary pressures. Research on swarm survival indicates that only one out of six lives to be one year old. Data indicate that queens are lost 25% of the time during mating. While drones are plentiful in the drone congregation area, the odds of any one drone mating with a queen is less than one in a thousand, meaning the majority of drones die of old age. For the queen the more drones she mates with increases hive genetic diversity. There may be sperm from 14 to 61 different drones stored in a queen's spermatheca which is expressed in all her daughters.

So it's a tough world being a honey bee. That is probably the benefit of the evolutionary process. Honey bees have survived for about one million years as cavity nesters. As a social unit, as a superorganism, bees have evolved behavioral methods rather than genetic mutations to deal with parasites and predators. Unlike a mosquito or house fly might develop a mutation to resists one disease, honey bees evolve behavioral systems to deal with multiple problems. For example, honey bee hygienic cell cleaning behavior has been shown to help minimize the impact of several bacteria, a fungus, a virus and a mite parasite. But the mechanism is so complex that all the genetic stars must align for the combination of two recessive genes to be present in the same queen at the same time. In this case, humans may help, armed with modern technology and careful breeding.

A few simple beekeeper management tools may prove to be enormous help to the honey bee colony located behind your garage. For colonies coming out of winter, a little feed, some room to grow and some simple management manipulation will make the difference between a bonus honey crop or a flock of swarms. A few extra frames of brood and the bees on them will make a new colony to replace the colony that died because the queen failed to mate last summer.

You may decide to leave honey on your colonies into the winter to ensure your new colonies survive. Along with

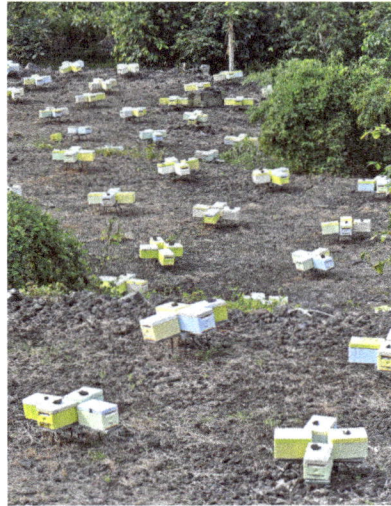

Mating nuclei like these in Hawaii ensure queens for hundreds of thousands of colonies worldwide.

that, new increase colonies can be tucked away along the edge of the apiary where young, mite-tolerant queens produce workers that clean out mite-infested brood cells, groom and bite mites off the body of other bees and reproduce so quickly that they mites cannot produce a fully-developed daughter before the bees emerge. Humans can help with that too.

Beekeeping is both animal husbandry and a carefully studied aspect of biological science. It also may be considered a partnership between two very different species of animals. These two species should and must be successful together, as their fates are be intertwined. For humans, help to the bees is rewarded when crops are pollinated, fruitful wildflowers support the ecosystem and surplus honey is stored in the comb. For the honey bees, a little help from humans can mean the species will thrive and succeed as long as humans are around.

Some are afraid honey bees will disappear off the face of the earth because of the actions of humans. Most beekeepers hope that, with their help, both species *Homo sapiens* and *Apis mellifera* (and *A. cerana*), will move forward together and thrive as two species in a mutually beneficial balance."

Greed versus Helping

When the nectar flow is over, often in the summer or fall, the beekeeper must decide to remove frames of honey, getting their 'share' of honey. If a beekeeper harvests all the honey found in the hive and feeds back lower priced high fructose corn syrup, the banker and the economist are happy—but only if the bee colonies live and do not need to be replaced. It is a choice of human greed versus animal care and welfare.

If the beekeeper takes out a few frames of honey for personal use and leaves the rest for the bees, it greatly reduces the need to feed the bees but gives the beekeeper a taste of honey. If the beekeeper then feeds the colony, it increases the risk of contamination of the remaining honey with sugar. The beekeeper is betting that the colony is going to be alive in the spring and will not need to be replaced. They are also betting that the colony has a low mite load based on monthly sampling and evidence that there are fewer than 2 mites per 100 bees in the young bee population.

Either way, where the winters are milder and there is a chance of a fall nectar flow, it is a gamble. Rather than play this game of chance, many commercial beekeepers move their bees in the late summer and fall to areas of Florida, Texas and other states where the winter is milder and there is a chance of a fall nectar flow. Large parts of Florida have nectar and pollen flows from August to November from two plants considered invasive and undesirable by ecologists and farmers. These are the Brazilian pepper bush and the melaleuca.[44] Beekeepers from northern states who seek sites where these plants are dominant by using the potential of a 2 to 4 mile radius as the basis of their selection may watch their bees fill the brood nest, produce honey and even raise drones in the fall of the season. This lets them produce queens to renew their colonies.

Small-scale and semi-professional beekeepers will trailer a load or two of bees to a warmer location to benefit from the late January-February red maple and willow bloom, two plants that are native species and predictable in their honey and pollen production.

There are many areas in southern locations in the United States where beekeepers locate their hives, avoid a northern winter, and benefit from an early spring and avoid the risk, the gamble, of wintering bees.

Now commercial beekeepers are building inside wintering facilities, some holding tens of thousands of colonies as a cost-effective means of overwintering. Keeping the bees under red-lights (which they see as dark) the practice prompts the hives to become broodless and consume less food. There is evidence that by keeping bees in a high CO_2 environment the bees remain alive, but in a stupor, while a large percentage of the varroa mites die. The beekeepers can then apply a mite treatment in the early spring, when the colonies are broodless, and then the hives are ready for a nectar flow with very few mites or viruses.

Science Helping Bees

Perhaps one reason to help bees, to keep bees, is to identify diseases and pests unique to them. We have discussed varroa mites. Now I want to discuss *Nosema*, microscopic pathogen of the midgut of the honey bee. The key benefit is to understand when we have a *Nosema* infection in our colonies. Once trained, any beekeepers can use a microscope to search for the spores.

For the small, backyard beekeeper, like myself, I believe the best option is to leave an over-supply of honey in the hive after the nectar flow is over in the fall. One example of this is leaving three deep eight- or ten-frame brood boxes on the hive, containing a large bee population of mostly younger, winter or fat bees that

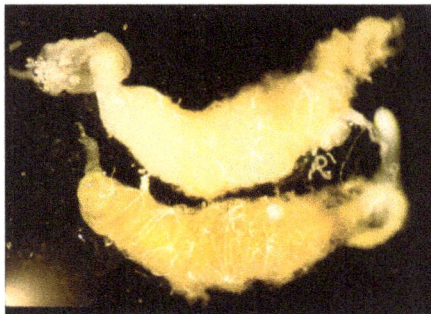

Midgut of worker bees. The top one is swollen with *Nosema*. File.

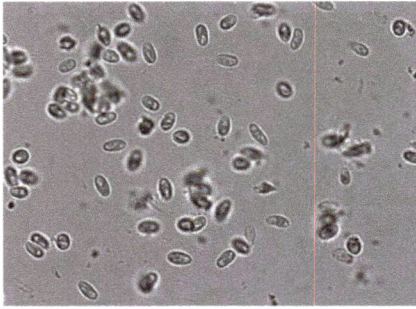

Spores of *Nosema*. File.

Fecal material associated with winter dysentery. It may or may not be associated with *Nosema* levels. Internet.

Inside wintering facility in North Dakota. Red light is not seen by bees. J. Miller.

have not yet raised brood. If a deep ten-frame box contains about 50 pounds or more of honey, it seems a smaller risk to take leaving it on for the winter.

Making nuclei and wintering them has worked for me. My five-frame nucleus boxes are behind the garage as they have been in past seasons during the summer and the winter. They serve as support colonies and are easier for me to handle.

Trying mite tolerant stock has worked for me. All year long. I also have an eight-frame, three-deep Saskatraz colony that came to me as a package. Bred for northern Canadian survival and production, a single colony is not a test, but it will tell me enough to see if I should make all my stock Saskatraz and learn to work with them in the future.

In past years I have used VSH, Minnesota Hygienic, Russian and Harbo VSH breeder queens. All produced beautiful open mated daughter queens and must be considered by all beekeepers.

At this point in my life I am just trying to remain open to new ideas and not forcing the bees into an old system of doing things.

7. SUSTAINABLE NATURAL BIODYNAMIC

A discussion

The primary goal of most beekeepers is to keep bee colonies alive and vigorous, but sustainable and natural beekeeping is no simple task. Contemporary beekeeping reflects over 9,000 years of honey robbing, biological investigation, and in the last few hundred years, experiencing a period of intense invention, innovation, and hard work to develop the kinds of hives and harvesting equipment we use today.

Beekeeping operations come in all shapes and sizes, including small scale, semi-professional, and commercial. In the United States, Canada, and Europe they all must observe laws requiring a standardization of hives for disease inspection. These standards include top-loading, movable-frame hives that honor the bee space[45] described in L.L. Langstroth's 1859 patent.

While Langstroth hives are used in most beekeeping techniques, there are other hive types that vary in the Langstroth concept and use different frames and box types. Elsewhere we have discussed other hive types. For now, let's focus on HOW beekeepers use their bees and equipment.

Most management practices are conservative and focus on the production of comb and liquid honey, bees for pollination, bees and queens for making increase colonies, and the production of package bee colonies. Each of these require specific management objectives and because nothing is simple, not everyone agrees with the everyone else's approach. Later we will review some of the methods needed to sustain and manage colonies using a natural set of practices.

Langstroth hive.

Contemporary Beekeeping

The majority of beekeepers in the United States and Canada use Langstroth hives, with standardized box dimension ranging from 3 to 12 frames, each with speciality depths. These hives may be made of wood, plastic or both. They contain frames and foundation for brood rearing and honey production.

The ten-frame deep Langstroth hive is the most commonly used hive in North America. Standardization allows for efficient manufacturing costs, making it possible to harvest honey with high-speed uncappers and motorized extractors.

In my backyard I use eight-frame deep hives for honey production and five-frame deep nuclei hives for queen rearing and as support colonies. While the boxes have different widths, the frames are the same, allowing me to move them inter-changeably between colonies. This allows me to equalize different strength hives and making new colonies. I successfully overwinter five-frame colonies in 2 to 4 hive bodies. The weight of each box is an issue for me, so I like that the eight-frame hives are about 20% lighter to lift and the five-frame nuclei are about half.

Location is critical in contemporary beekeeping. Small-scale and semi-professional beekeepers often keep their bees in one location throughout the year while commercial beekeepers are more likely to move their colonies back and forth between a home base and California (for almond pollination), various pollination contracts, and finally, a honey production area. If they have a queen rearing operation, this is in a warmer location for early or late season queen mating. Some commercial beekeepers operate from a northern state location for honey extraction in the summer and a southern or California location for buildup and almond pollination preparation.

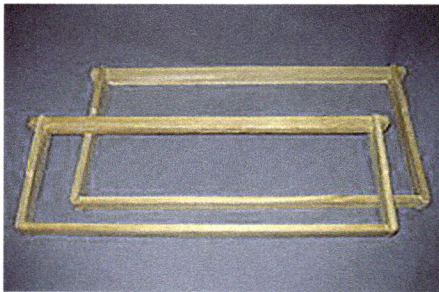

Deep and medium Langstroth frames.

Almonds are the single most important crop honey bees pollinate in the United States, and almond growers have a profound impact on commercial and semi-commercial beekeepers. Almond growers are riding a huge market wave, as almond nuts, milk and ground almonds are staples in my kitchen. Also, as the Chinese middle class grows and prospers, the demand for these protein-rich nuts continues to explode.

Almond flowers demand over two million colonies for adequate pollination.

Beekeepers are sharing that wave, benefiting from the huge pollination rentals and buildup of their colonies, but putting their bees and businesses on line in the process. The risks are enormous.

Mini mating nucleus with three small frames and about 1,000 bees. When made up it was given a queen cell and and sugar syrup.

Natural Beekeeping

The term "natural beekeeping" means different things to different beekeepers. To some, it means using 100 percent all-natural comb without any pre-embossed plastic or wax foundation, and to harvest the honey, the beekeeper must physically crush combs rather than using an extractor. Other people think that natural beekeeping prohibits using honeycombs that ever contained developing bees for honey storage. This, in despite of the fact that most bee colonies in bee trees move up and down as the year progresses, using brood comb for honey storage in an annual rotation. Then a few natural beekeepers determine their management for bee work on seasonal indicators like the phase of the moon, the summer solstice and other determinants. Finally, a few natural beekeepers base all their colony management on the writings and teaching of just one person.

A wide range of bee equipment is considered natural, although all are human-made. These include the Kenyan Top-Bar hives, long hives, Warré hives and hive types that allow for movable frames in agreement with apiary laws. In other parts of the world, outside the United States, Canada, Australia and most of Europe, a natural hive may be unmovable frame types: a hollow log, clay cylinder, woven box, and other, so-called primitive hives used where resources are limited. A bee skep would be an example of this type of hive, with its lack of movable frames makes it illegal to use in the United States and Canada. Simple

The frame is a standard wood Langstroth frame, but the wax started as a starter strip rather than a sheet of foundation. Bees are filling it with nectar.

Box 7. A Little Slice of Bio

I am writing this the year I turn 74, collecting observations and experiences made over a lifetime with bees and beekeepers. My beekeeper friends say they wish they knew what I have forgotten. Well, so do I! Here are some things I do remember.

On the family farm outside of sleepy Galesburg, Michigan, back when I was a toddler, bees were in our backyard, owned by a commercial beekeeper. Hearing the buzz each morning, feeling the sting on my bare feet, and savoring each drop of fresh honey were as much a part of growing up as spending a dime at the movies Friday night. It was the post-WWII era when jobs were not caused by war, food and gasoline were cheap, and money was always tight. My buddy Jeff and I spent the days in the fields and woods, checking in at home only to see what our moms were doing or if an apple pie needed a taste test. Or stuff warm chocolate chip cookies into our pockets. More often than not, we were shown the chores we had avoided.

In high school I raised sweet corn for a market in Kalamazoo and put Christmas trees up for sale at the farm. Dad and I cobbled together hutches from scrap lumber and hardware cloth where I raised rabbits that, with four brothers, we ended up eating. We had plenty of food from the garden and eggs and chickens from the coop. The freezer and the fruit cellar were always full. I grew up thinking we were pretty lucky. But we never took a vacation, ever. There was never enough money for the seven of us to travel.

Dad grew wonderful fruit on his eight-acre apple orchard. While the now-banned insecticide DDT was a standard part of his university-designed pest control plan, it did not hurt the bees as much as modern chemicals do. But it did affect them. I saw how DDT diminished the robin population and now know how much contamination DDT inflicted on us, our kids and our grand kids.

Surrounded by bees, butterflies, and beetles, I became interested in entomology and apiculture as a boy in 4-H. These efforts lead to scholarships and contacts at Michigan State University with whom I still speak with to this day. When I considered graduate school, I was offered an assistantship in entomology to work on honey bee pollination. By 1967, I became a professional apiculturist and first wrote articles about cucumber pollination. As the days went on,

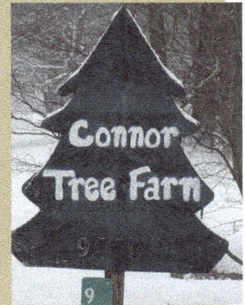

I learned about bee management from a more academic perspective and began speaking to beekeepers at bee club meetings.

Then there was extension work about bees at The Ohio State University[46] and setting up a bee-breeding program in Florida,[47] While raising two children as Mr. Mom, I acquired Wicwas Press from Roger and Mary Lou Morse in 1989. In 2007 I returned to Michigan and raised my own bees and queens on the family farm in Galesburg. Now the farm is sold, I have a few hives behind my urban garage in Kalamazoo where I'm writing from now.

Michigan State University apiary about 1969.

Most of my experiences incorporated a long-term, animal-friendly and highly sustainable approach to bee management. The concepts of keeping colonies alive I use ultimately trace back to the earliest human interactions with bees. The goal is to grow the number and size of colonies for maximum productivity when it is most advantageous to both the bees and beekeeper. Maximizing bee populations is essential for efficient crop pollination as well as creating the optimum number of forager bees for honey production.

woven wood and straw boxes are used in areas of limited resources with heavy scrap plastic bags or banana leaves covering the hive.

Some natural beekeepers use the Langstroth hive boxes and the top bars of frames, but they leave out side or bottom bars. They use starter strips to let the bees build their own "natural" combs on these top bars, allowing the bees to build the comb.

Biodynamic Beekeeping

As a kid, I was not aware that there were complicated educational theories on how teachers should teach and how my classmates and I should learn. I only knew that I needed to learn how to read, master the multiplication tables and pass geometry. If I sassed the teacher I was introduced to the flat side of her ruler or the oak chair in the principal's office. Educational theories behind the ruler were often lost on me, but the ruler's message was clearly received.

The same seems to be true of beekeeping. Once you get into beekeeping you find out there are groups of people with very strong theories and opinions. When you compare our actually learning process to the detailed theories and principles of beekeeping, there is quite often a huge disconnect! As a scientist, I was trained to use the scientific method as my guide to work bee colonies. A mathematician, on the other hand, might focus on the complex numerical relationships and geometry of the hive. A musician or artist might find inspiration and meaning from the sounds and beauty of the bees as they dance and gather food. Each of us find our own unique connection and conversation with bees.

There are some pretty variable concepts of beekeeping in the world. Commercial beekeepers rush to adapt any method they think will help them work efficiently and make or save money. Hobby beekeepers are often less interested in either efficiency or money and sometimes unintentionally harm their bees by spending hours making a hive inspection, stressing the bees and everyone within a hundred feet or more by excited bees. Not a sustainable practice.

In my slice of bio you can read a bit about how I grew up. I didn't say that while I was not an especially nervous kid, I usually was quiet and a little too serious, especially about schoolwork. Usually very shy, I took my position as the number two kid in a family of five boys pretty seriously. It was the role assigned to me and I played it well. Nobody taught me that I needed to get out of the way when number one and number three got into it. I was usually the kid picked to baby-sit numbers four and five.

The rest of the time I was hard to locate. Slow to read, I often hid in a quiet place to enjoy a pile of funny books (comics) and *Life* magazines, looking at the photos and cartoons. That's my explanation why the books I write are filled with lots of color pictures. Outside, I watched the transformation of caterpillars into butterflies or moths in an improvised empty package bee cage.

When I found and reviewed a list of principles of the Biodynamic Beekeepers,[48] I was initially afraid that most beekeepers were not familiar with their ideas and wondered if they should be, mainly because I did not agree with all of them. My decision to spend time to discuss them in this book was made with the knowledge that these ideas and concepts are frequently heard at beekeeping meetings, seen throughout the Internet, and taught at bee schools. They must be discussed.

Many of these ideas and teachings are of value to both the new beekeeper and the experienced hive owner. These are concepts every beekeeper should review as food for thought. Rather than ignore or deny the existence of these concepts, I feel it important that all beekeepers be aware of them and openly discuss them with seriousness. Since I have had a few decades of working hives and raising queens, I thought I would share some of my experiences and opinions. I'm hoping, like learning that the area of a circle is determined by its radius, some things need to be studied and mastered.

This is familiar territory. I grew up knowing about using natural methods to grow things. As a kid, a neighbor showed me her method of brewing "manure tea" to fertilize her garden. I helped her shovel cow manure and used bedding straw into a large wooden barrel and fill it with water. After a period of time, she removed some of the liquid to feed her vegetables and flowers. Her garden was always lush and verdant, with large tasty vegetables and stunningly beautiful flowers. She avoided bags of chemical fertilizers like Dad used to spread over our garden, although Dad never turned down a weed-free load of cow manure to spread on the garden.

Keeping bees alive is all about the management we use with our colonies and often seems to be a matter of common sense. Other systems are unique and radical, but intend to help the bees.

Natural biodynamic beekeeping projects this goal: "to minimize stress factors and allow bees to develop in accordance with their true nature." The promotional work available from biodynamic beekeepers suggests to me that they are reacting negatively to the view of contemporary commercial beekeepers apparently abusing these insects during the course of their production of new colonies, transportation, queen rearing or honey harvesting.

Biodynamic beekeepers have an interesting list of do's and don'ts they follow. In this chapter I devote careful thought to evaluating practices important to all beekeepers—especially for those trying to keep their hives alive throughout the year. Let's study biodynamic beekeeping and review the advantages and disadvantages of its teachings.

Here is a list of rules or guidelines of Biodynamic Beekeeping that I have organized into three sections: Going Natural, Queens and Swarms, and Minimal Chemicals.

GOING NATURAL

Use of all-natural materials—beeswax comb, wood, straw and other materials

Natural Comb

Natural combs are used rather than foundation

Advantages	Disadvantages
Bees are excellent natural comb builders.	Bees have their own ideas about comb building and the results are often impossible to inspect.
Less chemical contamination potential.	Bees built combs are fragile and break with handling.

What kind of comb should we provide our bees? Should we use the machined sheets of beeswax or plastic embossed with a hexagon pattern or let the bees build their own comb?

Bees do a fine job of building comb, having done it for millions of years, creating irregular patterns, attachments to other combs, communication and travel holes and natural bridges and braces. They obey the space between combs, the bee space, but maybe not the orientation of the combs. At least not to human standards. Humans want to improve on what bees do. But the bees do not always build it exactly the way humans want it.

Any comb bees build themselves is considered natural. The use of pre-embossed foundation made of beeswax, plastic, or another substance (aluminium was once used to replace beeswax during war years, when beeswax was essential to military waterproofing) does not fit into the "natural" category. Most of my beekeeping experience has been based on human-manufactured honey comb foundation because it is an efficient system for bees building new honeycomb and labor-saving for beekeepers. It provides the bees with a starting template and a bit of wax to work with, helping the bees by doing some of the work for them.

With manufactured comb, a predetermined cell size is embossed in the wax or plastic foundation, but when bees build their own comb, the cell size is probably the size that the bees will most effectively utilize.

There is a contemporary debate about cell size related to varroa mite control. There are many advocates who want bees to be raised in smaller cells because they believe these cells should generate fewer varroa mites. There is no guarantee that small cells will reduce the development of varroa mites, and the science on small-cell beekeeping shows that it does not provide the desired advantage.

But to be fair I would like to see more work done on this subject, especially on the aspect of small cells resulting in faster brood development. The evolutionary host of varroa, *Apis cerana*, has decreased worker bee developmental time and the mites only reproduce on drone brood, which takes longer to develop. Certain

Comb built in a wall by a swarm. C. Hubbard

New comb constructed in an empty space in the box.

Natural comb built on a paint stirrer.

Thin wax as starter strip

Wired wax foundation.

Exposed black plastic comb with larvae.

African races of *Apis mellifera* also have faster developmental time. Is this cause and effect of cell size, or just a correlation?

Small cells go against a century-long industry movement to develop larger bees. Larger worker cells and larger bees are considered desirable among many beekeepers. This is true with different races (sub-species) of bees and was the root of the Starline and Midnite hybrid bee program I contributed to from 1976 to 1980.

That argument is that larger bees collect larger nectar crops and store honey more efficiently than smaller bees. There was a push to use larger bees in North America starting with the importation of Italian queens and part of the arc of progression toward 'modern beekeeping' in North America since the 1850s.

Some bee colonies populated with larger bees demonstrated hybrid vigor (heterosis), which occurs when unrelated races or inbred lines are crossed. By producing hybrid bees from intentionally inbred lines with concentrated genes desirable in the final hybrid, we produced bees with heterosis, and increased bee size was part of that science.

A big advantage of natural comb is the significant reduction of the amount of pesticides and antibiotics present in the hive.

These compounds have an affinity for the wax. The wax comb is compared to the human liver except for one important detail. While comb stores poisons encountered by the bees, there is no mechanism to detoxify or remove these poisons from beeswax comb the way the human liver does. As a result, these compounds persist for years in the brood frames and throughout the colony. This is one reason many Langstroth hive beekeepers replace their combs every few years.

I doubt there are many places in North America or Europe where bee colonies are pesticide free, picked up from the environment or added to the beeswax as beekeepers treat for mites and diseases. Government tolerances says their levels are sublethal, conflicting with wax-using industries wanting their products to be poison free.

Natural beekeepers often use other methods, softer chemicals, drone trapping and other varroa control methods as a management tool against *Varroa destructor* and American foulbrood. These are excellent steps to use to reduce comb contamination levels. Even if firm believers of natural beekeeping do not intentionally expose the natural combs in their hives to pesticides and antibiotics, these compounds still end up in their hives.

Are the current levels of pesticides in beeswax endangering honey bee health? Human health? The question is indirectly answered by the Roman Catholic Church and cosmetic industry. These are two large users of pure beeswax. Both seek minimally processed *Apis mellifera* beeswax from countries where pesticides are not used because of their high expense. The Catholic Church has long advocated the practice of using 51% pure beeswax candles during services, spiritually associating their beliefs of the virgin birth with the unmated worker bees that make the wax. The practical aspect is that these are superior candles, creating less soot and odor when burned. The cosmetic trade, which deals with products and compounds that come in direct contact with human skin, seek the cleanest wax they can find.

Some countries in Africa and South America, produce much of the "pesticide free" beeswax used by the cosmetic trade.

Box 8. Plastic Comb and Equipment

Drawn-out black plastic comb with eggs.

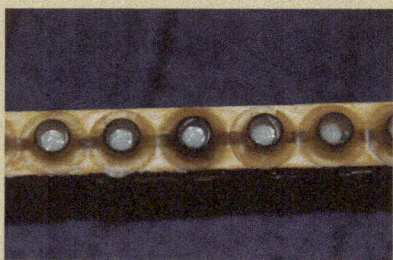
Plastic queen cups containing larvae.

Queen cell in plastic queen cup showing an abundance of royal jelly.

Plastic honeycombs and hive boxes come with two issues for beekeepers to consider, the first being that most plastic will outlive both the bees and the beekeeper. The plastic combs will likely be burned or put into landfills. We must find—and use—a method of recycling plastic combs.[50]

Second, most plastic comb foundation is sprayed with a thin layer of contaminated beeswax. The wax encourages the bees to 'draw out' the comb adding more wax to build comb. Uncontaminated wax is difficult to find.

The beekeeping industry could have a policy of recycling old plastic combs by returning them to the dealer who initially sold the item, like in states with cash deposits on carbonated beverage containers.

Also, if a new beekeeper starts out with a few combs, plastic or wax, so the bee colony starts development and growth, and then converts to natural comb, they may find a suitable compromise.

By using all-natural starter strips in Langstroth, Kenyan Top-bar, or Warré hives, the beekeeper accepts that a certain number of combs that will not be perfect. Some must be cut and pieced into a functional comb or put into the melting pot and converted into beeswax blocks for candles, cosmetics and other products made by the beekeeper.

If the beekeeper uses a honey extractor (which may not be needed if the bees produce good quality natural honeycomb sold as section, cut comb or chunk honey), a few combs may be broken. Retrofit the comb in the frames or put the broken bits into the melting pot. Accept this loss as a small cost of your decision to produce natural combs.

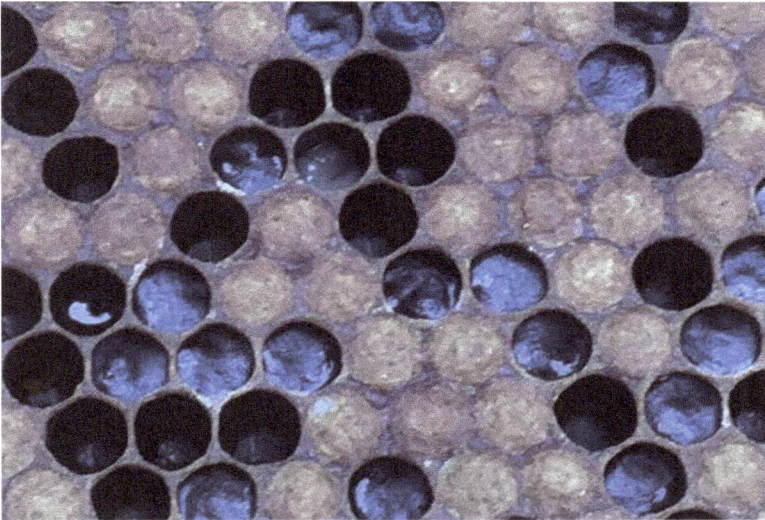

Bees have their own methods of dealing with contamination. Here we have pollen cells sealed with propolis, perhaps sealing in contaminants. J. Winer.

Crock pot filled with melting comb, a flameless way to melt wax. The dark color is from layers of cocoons, propolis and age-darkened wax. This is called slumgum.

183

The amounts of these pesticide-free wax supplies are rapidly shrinking.[51]

Minimally tainted beeswax, if attainable, should be used in the production of beekeeping comb. Save fresh capping wax—the thin, new wax from the top of the honey comb—and use it for comb production. I feel that the use of natural comb is often the best answer for small-scale and sustainable backyard beekeepers and should be carefully considered by all beekeepers.

Made of Natural Materials

Beehives must be made of all-natural materials, such as wood, straw, or clay

Advantages	Disadvantages
Bees show flexibility about their nest type.	Durability is an issue.
Hollow logs, woven baskets and clay are less inexpensive.	Hard to keep top-loading, movable frames.
Do the bees care? Probably not.	Treated wood is toxic to bees.

As in our discussion of comb, the homes we put bee colonies into are also part of the biodynamic beekeeping conversation.

With a strong anti-plastic argument, the biodynamic doctrine bans anything not naturally made. I prefer cotton over polyester but have worn both. While cotton holds moisture and body odors, it is from plant fiber while polyester is produced from crude oil. So, cotton use should win the environmental, sustainability argument against using polyester.

The same is true in the type of bee hive you use. I have several polystyrene nuclei boxes and like them. I find that the bees winter better in these than when bees are kept in wooden hives. Commercial queen producers are converting to polystyrene mating nucleus boxes because they get better results—better queens and more of them—from plastic boxes when compared to wood.

While I have not kept bees in skeps or woven hives, clay cylinders or other natural materials, I know they work very well for beekeepers in many countries. When European humans entered North America, wood was abundant and easily crafted into a box faster and easier than weaving a basket. Skeps were no longer used in New England by 1820.[52]

Polystyrene mating nucleus for five deep Langstroth frames. Warm syrup may be fed into the hive during the winter, poured through the entrance.

Look at what the bees use! When given a choice, bee swarms use many non-natural human-made cavities (mailboxes, bird houses, empty grills, water meter enclosures) with success. During scout bee inspection and the decision to enter a cavity, the material used to create the colony's future home does not seem to be as important to the bees as the size, volume, lack of repellant odors, and soundness of the cavity.

All plastics should require a recycling or disposal plan. Few, if any, beekeeping hive parts have them. They should all be marked with a recycling symbol.

My friend Bo Sterk started working with young beekeepers in the mountains of Haiti after a hurricane and earthquake devastated the island. He found that a few hives of bees produced enough honey to fund their college education. These young people now teach in their mountain villages. Since there is little or no money for bee equipment, hives were created out of hollowed out logs, woven hives, and salvaged plywood sheets. In fact, Bo instructs them to build Kenyan Top-bar hives from plywood. Abundant banana leaves serve as the covers, keeping the bees dry. The same leaves wrap honey comb harvested from the hives. Bo has experimented with various types of hive bodies, even half an oil barrel. When you have nothing, having something is much better.

185

U: Young Haitians with Kenyan Top-bar hive made from a sheet of plywood. B. Sterk.

M: Hollowed Haitian log with colony inside. Banana leaves cover the end. An enamel pan is the smoke bucket. B. Sterk.

L: This half metal barrel serves as a hive in Florida with top-bar frames. B. Sterk.

Hand-hollowed log holding a colony in Kenya. The log hives hang from trees. S. Repasky.

Combs are harvested by removing the wood disk that closes the entrance. S. Repasky.

Almost all of the equipment I have used for about a half a century, has been made of wood—pine or cypress in most cases. This is considered a renewable resource suitable for the biodynamic concept.

For years we have known that propolis (resin collected from trees supplemented with beeswax) helps bees manage moisture, bacteria and parasites. Now we know it also minimize viruses. Bees use propolis in some plastic and wood hives. Some beekeepers have picked up on the pro-propolis research and leave the inside of their hives rough to stimulate the bees to instinctively coat the interior with a layer of propolis. Or they paint the inside of the hives with a propolis varnish. This has been shown by researchers at the University of Minnesota to increase the health of the bees' immune systems.

A layer of propolis at the top of cells provides strength and antibiotic properties.

No Pollen Substitutes
Pollen substitutes are prohibited

Advantages	Disadvantages
Pollen feeding is not always necessary.	Pollen feeding is more expensive and labor intensive.
Bees collect a diversity of natural pollens.	Pollen feeding can spread AFB, chalk brood and viruses.

Disease- and chemical-free natural pollen is the best protein source for bee colonies, especially if the bees eating the pollen are from the colony that collected it. Avoid pollen contaminated with pesticides or antibiotics—bee colonies perform best when fed a diet of naturally collected pollen. Unfortunately, few beekeepers collect pollen from their own hives and feed it back when needed, and it is impossible to tell if pollen is chemical free without testing.

In areas with diverse plant communities, colonies collect a mixture of plant pollens and do not often need supplemental protein feeding by beekeepers. Protein feeding becomes necessary when pollen diversity is limited, as bees benefit from a nutritional mixture of pollen sources. In agricultural monocultures, the composition of the natural and human-planted may be good or bad for the bees, depending on the crop. In almonds, the bloom yields almost 100% almond pollen, but it is nutritionally complete.

The same thing can happen in nature: my bees benefit from a huge flow from the aster flowers just after the goldenrods are finished for the season. Potentially floral monocultures, almond and aster pollens are protein rich and no supplemental feeding should be required. Unfortunately, not all plants produce much pollen. Some plants fail to produce pollen under certain conditions. A new study of pollens produced in urban environments show a remarkable diversity of pollens.[53]

Some beekeepers collect natural pollen using special pollen traps when there is an abundance of pollen being collected by the bees. Others make sure the frames in their colonies contain freshly stored cells of pollen, converted to bee bread by the bees' addition of various natural microbes that facilitate lactose fermentation, and made available to the bees for food. Bee bread is better for young

L: Worker removing material from a protein and sugar patty using her proboscis.
R: Many bees join in to remove the stimulating food.

bee development than dry pollen, keeps its nutritional value longer, and is nutritionally better than any protein substitute.

Beekeepers who collect their own pollen from their own colonies must carefully inspect their hives and the collected pollen to ensure there are no pieces of chalk brood mummies, American foulbrood scale or history of high levels of viruses (such as deformed wing virus) present in the hive. NEVER use pollen from other beekeepers as the risks are much too high.[54]

Human-made protein mixes, called pollen substitutes, fill the pages of bee supply distributor catalogues. These substitutes do not contain natural bee pollen because of its cost and risk of disease spread. These mixes are often released after prolonged testing by research laboratories and beekeepers.

I used winter patties with success. These low-protein, high-sugar content patties ensure that overwintering colonies have a trickle of protein available during the late winter and early spring as their brood rearing initiates. The real benefit is the high sugar content, which stimulates consumption and brood rearing.

By feeding my bees the colony is alive and growing a new bee population before the natural sources of pollen (skunk cabbage, red maple, alders and willows) can be collected by the bees and the new season has started. This is good for the bees. I stop feeding as soon as natural food is available as the season develops. These winter patties are primarily beneficial because of the high amount of carbohydrate present, not just the proteins.

There are a number of reasons why beekeepers avoid using pollen substitutes. Some avoid using mixtures of sugars and protein sources that include animal-based proteins (milk solids for example), choosing instead to feed an all-vegetable based diet.

It is not necessary to feed when the bees have plenty of food and feeding would be wasteful. Feed in apiary locations that are nutritional wastelands, lacking natural late-summer and fall pollen production. If your colonies cannot collect pollen in the fall from goldenrod and aster, because of drought perhaps, they have a greater probability of dying over the winter months. Protein supplements, thus, become necessary.

Protein and pollen patties are highly attractive to small hive beetles, and in some cases, wax moths. Feed small amounts, what the bees can consume in three days to minimize beetle development risk. After three days the beetle larvae will hatch and soon coat the frames with slime.

Sell Honey in Glass or Metal

Honey may be transported in containers made of artificial materials but must be decanted into containers of glass or metal for retail sale

Advantages	Disadvantages
Glass and metal are good packaging materials.	Plastic lasts a long time and must be recycled.
New packaging for honey is necessary.	The market favorite, the honey bear, must die.
No chance of chemicals leeching	Plastic is sometimes iffy to reuse.
Many reuse and recycle metal and glass containers.	Aluminium reacts with the acid in honey.

No plastic containers for honey! Kill the plastic honey bear! While it is not a part of keeping bees alive, abstaining from the use of plastics is part of the discussion about keeping beekeepers in business and global sustainability. There is a lot of hysteria about chemicals contaminating honey from certain plastics. I am in the category of "when in doubt, why take the risk." Metal cans and glass jars are fine ways to store and sell honey, but not aluminium, which chemically reacts to the acid in honey. But why don't we sell our honey in the comb, wrapped in a piece of natural cloth, a beeswax-coated cardboard box, fabric, or a banana leaf? I'm serious!

As a species, humans must come to grips with the contamination of the planet with plastics. For this reason alone, I use glass containers for my honey.

Fresh honey from my bees in a glass container. Bubbles float up to clarify.

Back on the farm, as a child growing up in the middle of the last century, we were one of "those" families that canned and froze a lot of our food—most of which we raised. As an adult I can my own tomatoes and peaches because I like how they taste. Maybe, just maybe, there is a niche market for your honey out there, honey produced within 100 miles, that you can sell in natural packaging and in the comb. I'm thinking of a beeswax-coated card stock-or cloth wrap that can be recycled or reused.

SWARMS AND QUEENS

Swarming[55]

Swarming Is Recognized As The Natural Form of Colony Reproduction

Advantages	Disadvantages
Swarms are an excellent way to get new colonies.	Swarms can reduce bee and honey production.
New swarms occupy a key niche in the ecosystem.	Swarm colonies increase the risk of queen failure.
Established swarms provide drone production	Potential of making splits and using packages reduced.
Help develop natural mite tolerance in isolated regions.	Swarms carry potential diseases, parasites and viruses.

One of the arguments from biodynamic beekeepers is to rely entirely on swarming as the sole method to make new colonies. I have an issue or two with this, so let me explain why and how:

Certainly, swarming is, of course, nature's way of producing new colonies—totally biological, instinctive and essential for species survival. Unlike yellow jacket or bumble bee queens that establish nests with just one reproductive female, honey bees have evolved the behavior of splitting strong colonies into the parent colony and one or more daughter swarms.

Prime swarms contain an average 12,000 bees, the old queen from the mother colony (usually), and a few hundred drones that were caught up in the departure of so many bees from the entrance of the hive. The bees use their amazing dance language to vote on a new nest site some distance from the mother colony and fly to it en masse to quickly build a large bee population and store abundant reserves to become permanently established.

Because all queens in all colonies carry the genetic instinct to swarm. I do not consider honey bees to be domesticated. That word implies an animal is tamed like a care-dependent pet or farm animal whose food supply and often reproductive cycle are controlled by humans. Bees do not need human intervention to survive and have evolved fiercely independent reproductive systems.

It is the beekeeper that enters the apiary with a veil, smoker and hive tool for protection and offers the bees food and medication for the bees in return from their human servants. The bees are free to leave the hive at any time, unlike most pets and farm animals, and the queens are free to mate "with whatever drone I encounter."[56]

Queen cell under construction. The larva is inside, floating on a bed or royal jelly.

Cups or early cells in the bottom of the frame position typically associated with swarming.

Low hanging swarm.   A small swarm indicates it
is an afterswarm.

Beekeepers have a wide range of methods available when making new hives. They may collect swarms, but are dependent on the bees' timing. When some Michigan General Motors workers were let go from their jobs after the 2008 economic crisis, they chose to hang out in their apiaries to capture swarms. Their swarm collection behavior fit the biodynamic paradigm. Few of us have the luxury of time to spend with their colonies.

Packages are often called artificial swarms. Unlike natural swarms, because of the way package colonies are produced, the bees and the queen are rarely related.[57] Instead, a foreign queen is introduced to bees shaken from many colonies. Then, the bees are shipped to a new location, perhaps thousands of miles away and into a new flora biotype.

Swarms are local. They did not depend upon fossil fuel to arrive at their new home. Swarms sometimes appear superior to packages of bees, but are not without problems or risks.

Whenever I use package bees, I add a frame of sealed and emerging worker brood, if I have it. When done properly, adding this brood provides bees of ages that are not present in a package colony's age demographic. Adding a frame or two of stored bee bread and honey to the package benefits the growth of the hive, reducing—if not eliminating—the potential for the colony to abscond, when the bees fly out of the hive and find a new home. Brood and frames of honey help package bees become established.

Up front, package bees are an efficient and traditionally cost-effective method of starting bee colonies, replacing winter losses and growing an operation. Yet in the hands of inexperienced beekeepers, package use leads to overwhelming mistakes, errors of judgment and total failure. When losses and queen failures are factored into their final cost, they are often prove to be a very expensive way to start a hive of bees.

When I have overwintered colonies, my favorite method of developing new colonies is by splitting a hive. During splitting frames of worker brood are moved from the mother colony to the daughter unit, and a queen introduced. I start new colonies by splitting a large hive into two or more colonies. Each hive is a miniature version of the original, containing bees and comb, stored bee bread and honey, and a fraction of the resources found in a full-sized colony.

I use and promote a method described by Gilbert Doolittle over 100 years ago.[58] All the bees are shaken or brushed off the combs.[59] Bee-free frames of brood are placed over a strong hive with a queen excluder in between to separate the two units. This system is explained in detail in my book, *Increase Essentials* as the Doolittle Increase method. The bee-less brood's pheromones stimulate the nurse bees of a colony to move upward and through the queen excluder to reach the brood to care for it. Done properly, there is no crushing or mangling of bees, only the peaceful reunion of young nurse bees to the next generation of the species. I add frames of pollen and honey, food and a new queen, often a swarm cell from the same colony. In a few hours the bees may be moved off the mother hive to a new location so the increase hive may grow and prosper. Because the colony lacks forager bees, I reduce the entrance to conserve heat and provide protection from robber bees from other colonies.

So, in response to the biodynamic argument that swarming is a natural method of colony reproduction, I totally agree. But it is not the only method. There are other ways to reproduce colonies efficiently while causing little harm to the bee. For new and small-scale beekeepers, the Doolittle Increase method is that way.

No Wing Clipping, Try Marking
Clipping of queen's wings is prohibited

Advantages	Disadvantages
Queens are not handled and wings cut.	Unclipped queens are difficult to confirm identity.
Marking is an alternative.	There is no record of queen replacement.

Upon swarming, the mother queen will often leave with the swarm, often considered a loss to the beekeeper. Thinking it will prevent her from flying out of the hive during the swarming process, some beekeepers clip a queen's wing. Unfortunately, wing clipping does not prevent swarming. It only delays the process by a few days, long enough for a new daughter queen to emerge with fully intact wings who will then fly away with the swarm.

I clip a queen's wing for one reason: to mark her for future identification. I want to make sure that the queen I have in front of me today is the same one that I observed a month ago. If I clip a queen's wing, I take off one third of the length of the wing so as not to interfere with her balance. In a queen breeding program, I only clipped the wings of queens with special genetic traits, leaving production queens[60] unclipped. All of the queens in an operation were marked with a small dot of paint on their thoraxes.

There are many good reasons not to clip a queen. Unskilled beekeepers may do the job improperly; a queen with one wing completely removed can't fly and may walk off balance. Worker bees may replace her, and you lose your investment.

Removing part of a queen's wing is the only guaranteed method of identification, but comes at some risk. It generally does not speed hive inspections.

Marking a queen, however, speeds queen finding and hive inspections. This is a good thing for the bees as it keeps well-intentioned beekeepers from spreading boxes and frames all over the yard while searching for a queen. That is not good for the

Do you need to clip and mark a queen? Marking helps find a queen but clipping is the only way to make sure the queen you find is the one you last saw!

bee colony as it will suffer loss of food gathering, brood feeding and even queen stability as a result. Leaving a hive open that long can also prompt robbing. Many queens are killed or lost when unskilled beekeepers work their hives.

To mark a queen, a very small amount of paint is gently placed on her thorax. Any queen may be marked, carefully, with a tiny drop of paint or with some glue and a special numbered metal disk. Only a small amount of paint is needed, so skill is essential. If too much paint is applied, it may result in the material spreading over the entire thorax and head of the bee. If this happens, you have just killed your queen!

Become proficient before marking a queen by practicing on drones. Some of these markers flow a lot of ink and must be carefully checked before use.

If you clip a wing of the queen, take off no more than one half the length. Do not accidentally remove a leg!.

197

To reduce the amount of paint use a straightened paper clip (a small one) just touched to the inside lid of a model paint jar. Some model paints and paint pens in use may be toxic to the queen, so test the material on drones before you mark a large number of valuable queens. I have had many paint pens fail, leaking paint all over the queen's body and killing her. So be careful.

Unfortunately, the paint on a marked queen can wear off or be chewed off by the workers. Only by removing a part of a queen's wing will you guarantee her identification.

No Queen Replacement
Regular and systematic queen replacement is prohibited

Advantages	Disadvantages
Colonies usually produce superior queens.	Timing is up to the bee most of the time.
Sister queens fight and eliminate weaker sisters.	Old queens decline before being replaced.
Swarm and supersedure queens are perfect for increase.	Natural re-queening may lower productivity.
Requeening causes a break in the varroa reproductive cycle.	

Biodynamic beekeepers prohibit 'regular and systematic' queen replacement. It is true that most beekeepers are obsessed about queens and their replacement. Yet in many commercial operations, one- or two-year queen replacement schedules keep the entire apiary productive and at maximum efficiency. This is often measured as the average size of the colony rented for crop pollination, the pounds of bees shaken for package bee production, or the weight of the honey the average colony produces.

Migratory beekeepers routinely split their colonies by making increase units during the spring or in the late summer or fall after pollination contracts are completed and honey production is done for the season. Other beekeepers located in a Sunbelt state do this

A nice queen cell can be used to make a new colony.

Plastic cage helps in the queen introduction process.

in the late winter and early spring. As this happens the colonies get a brood break that helps defend them against varroa as the colony rebuilds on new, less-contaminated combs. Hopefully the beekeeper has installed a young, well-mated, mite-tolerant queen.

If beekeepers send colonies to almond pollination in California (as nearly all commercial beekeepers do), they seek to maximize colony populations in order to receive the highest rental fees when colonies are inspected and graded by third-party inspectors.

If a beekeeper chooses and does not care about apiary efficiency and increased average queen productivity, regular queen replacement is not a high priority. This is usually a passive decision, one not necessarily based on principles. It is the practice I use in my back yard as I observe how long certain queens last.

Yet, I have overheard large commercial beekeepers, folks who make millions of dollars each year selling queens to other beekeepers, admire how much they like a naturally produced queen; a queen from a swarm hive or a supersedure queen. That says that the biodynamic beekeepers may be onto something important—honey bee colonies are better than humans when it comes to the time and place of producing and mating queens.

Of course, most beekeepers find themselves involved in emergency queen replacement when a queen suddenly dies, or has failed to mate successfully. Because it is difficult to avoid queen replacement altogether, my instinct is to encourage most beekeepers to adapt a regular queen replacement program.

I do this by using natural, bee-produced queen cells for making increase nuclei. It means that I must time my increase production to coordinate with swarm and supersedure cell production. Whole frames of brood with attached cells may be moved when making increase. Use care to prevent damage to the cells. Use this system when using queens purchased from locally produced, mite-tolerant stock, as their daughter's drones will also be of the desired stock because they carry only their mother's genes.

No Instrumentally Inseminated Queens

'Artificial' insemination is not used. Instead queens are allowed to fly free to mate.

Advantages	Disadvantages
There are alternative plans to bee breeding.	Rejects the concept of rapid stock improvement.
Isolated mating is possible on islands and remote regions.	Few beekeepers have isolated mating locations.

My mentor and colleague at The Ohio State University was Dr. Walter Rothenbuhler. A global proponent for the proper use of instrumental insemination, he had a unique appointment in three departments: entomology, zoology and genetics. He was responsible for the research on the recessive genes that are seen in the hygienic behavior of bees, a tool used currently against varroa mites, American foulbrood, and three other bee diseases.

Growing up in southern Ohio, Walter had a carefully thought-out assessment of the insemination device he used to conduct his research. He said, "Instrumental insemination is like using a high-powered rifle—you have to know how to use it."

After OSU, I established and ran a Florida company that used instrumental insemination to produce thousands of instrumentally inseminated (II) queens. By doing so, we were clearly going against the biodynamic doctrine.

This experience taught me a great deal, especially how little we humans know about honey bee mating biology. It also taught me the importance of a skilfully run instrumental insemination program, something most beekeepers do not have the motivation or technical ability to accomplish. This is not a criticism as much as it is accepting reality.

Beekeepers were overwhelmed and ill-prepared for the use of instrumentally inseminated queens. They are special, often fragile creatures that meet our objective of controlled mating, something that is nearly impossible unless you have mating isolation, like an island somewhere with hundreds of known colonies producing drones of the proper genetic stock.

Stock improvement using instrumental insemination is and has been a valuable contribution to our goal of keeping bees alive and thriving. Because of the multiple mating of each honey bee queen, it is a solution to the problem of mating accuracy. To duplicate the accuracy of mating we had with the Genetic System/ Dadant program, we would have needed 13 isolated mating locations.[61] Where are there 13 islands where we can mate queens in isolation—one each for our 13 inbred lines?

We know that honey bee queens mate with multiple drones who all die after copulation, and whose sperm is stored in the queen's body for many months or years—as long as she lives. Queens do not mate again after their first few weeks of adult life. Honey bees use unique drone congregation areas (DCAs) where young queens fly to find about ten thousand drones from surrounding colonies awaiting her arrival.

The total mating process is often done in a few minutes. Queens mate with 14 to 61 drones, based on genetic testing of the sperm stored in the queen's storage structure, the spermatheca. I don't know of any evidence of queen's avoiding genetically related drones. Mating is a rapid-fire reproductive event that ensures each queen is fully loaded with sperm from as many drones as possible. It takes about two seconds for the first drone is pushed out of the way and falls to the ground as the second drone mounts the queen and enters her with his endophallus.

Estimates indicate that a single DCA involves about 15,000 drones.[62] At any point in time, about one third of these drones are flying to or from their hive for refueling with honey in their home hive so they can continue patrolling for young queens, leaving 10,000 drones patrolling in the DCA.

If a strong spring or summer colony has 500 mature drones flying in and out to the DCA on any given day, it means that the environment must have at least 30 vigorous colonies within flight distance to support these 15,000 drones. My math may be a bit shaky, but if colonies have low drone numbers, the bees shut down some DCAs (no, we don't know how), and remaining drones join other DCAs nearby. If an area has intensive beekeeping, there may a choice of the DCAs the young queen may visit to mate.

While I ran the Dadant Starline bee breeding program in Florida our business model was also to mass produce II queens for general use by beekeepers in production hives. The hybrid was a three-line combination called the Cale 876. While the program did not reach our financial goals (largely because of the cost of hiring humans and the unpredictable outcome of the II queens) the program had enormous benefits we need to review.

These queens were best suited as grafting mothers or mothers of naturally mated daughter queens, as the cost of production

Dorothey Morgan instrumentally inseminating queens using a newly designed Instrumental Insemination device at the Purdue Bee Lab. D. Morgan.

of each II queen was greater than open-mated queens. Because the daughters produced very successful colonies many of our customers quickly figured out that these queens were best used as breeder queens—mothers of naturally mated queens in any sized operation that would produce desirable and productive colonies. Because of the haploid-diploid methods of sex determination, the daughter queens could be open mated and still produce pure drones of the desired stock, using a simple but demanding program of drone saturation that made it possible to saturate an area with selected drones. I call these target drones, the male bees you want your young female queen to encounter and mate.

Today, I fully support the use of instrumentally inseminated queens as a means to reach a specific goal, meaning that you must be carefully trained and master the methodology. This echoes Professor Rothenbuhler's statement that you must know how to use the instrumental insemination technology.

II queens are a means of producing daughters that carry traits that benefit both the bee and the beekeeper. Since most beekeepers feel that the current key area of focus should be on control of varroa mites, there is a strong argument to use instrumental insemination as a means of collecting and maintaining desirable genetic traits that will can be transmitted to daughter queens that

Cage holding queen pheromone from helium balloons. In search of a drone congregation area (DCA).

end up mating naturally with a wide cadre of drones, creating a *superorganism* of mixed worker subgroups (each with their own father drone), that combine to form a vigorous colony, but one that has behaviors for disease resistance and of mite tolerance.

Another important note: Instrumental insemination allows for the preservation of sperm in gene banks, frozen specimens that may be used in the future. This could be an important concern if a particular line or race is eliminated.

Today, the II process has yielded several strains of bees that have improved bee health and mite tolerance, especially by using the hygienic behavior, grooming behavior, pollen foraging and overall genetic diversity. Outside the research laboratory most of these instrumentally inseminated queens are used to produce daughter queens that carry beneficial traits when they open mate with random drones.

No Grafting

Grafting of larvae to produce queens is prohibited.

Advantages	Disadvantages
Colonies usually produce superior queens.	No non-grafting system competes poorly with large queen cell production.
Non-grafting queen rearing systems exist.	Not all beekeepers can see larvae small enough to graft.
Swarm and supersedure queens are perfect for increase.	

Larval transfer from the worker cell to the queen cup produces large numbers of desirable virgin queens at low cost and high efficiency. Done properly, the queen quality is as good as anything the bees do on their own.

When Gilbert Doolittle published his *Scientific Queen Rearing* in 1889, he had spent years reviewing the challenges beekeepers faced in producing queens from specific queens. He gathered the developments of other beekeepers into one system that allowed the transferal of honey bee larvae of a young age into a wax cup that the bees expanded into a queen cell.

Biodynamic beekeeping prohibits the use of larval transfer or 'grafting' of newly hatched bee larvae into queen cups that are placed into a queenless colony for queen cell production. Why they advocate this I am not sure. True, the process is not easy to learn. Perhaps they are afraid they will hurt the larvae when they move it. That can and does happen.

There are other methods of producing queen cells that do not require grafting of the larvae. Methods by C.C. Miller, Jay Smith and the process of frame notching and cutting strips of brood are available for queen producer to avoid the grafting technique.

Non-grafting methods are appropriate for small scale queen producers. Few beekeepers actually need to produce several hundred or many thousands of queen cells a season. Creating queenless colonies was a method Langstroth used very successfully to harvest ripe queen cells by periodically dequeening colonies on a rotating schedule. Langstroth's dequeening method is not suitable for large-scale queen production because of the number of colonies it would take out of honey production and the low numbers of cells that result. Is the act of transferring a larva so horrible that it should be banned? It is a pretty insignificant process compared to what certain humans go through to achieve a

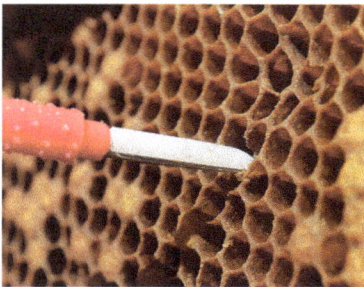

Transferring larva from brood comb to a queen cell.

Worker feeding a queen cell, showing abundant royal jelly in cups.

pregnancy. Are we to ban in-vitro fertilization and deny the joy of parenthood to so many families?

One may argue that naturally made swarm and supersedure cells are better than cells produced by a queen rearing operation—grafted or not. Of note, my use of the Doolittle increase method of making new colonies allows me to use swarm and supersedure cells from strong, desirable colonies and completely bypass the grafting process.

The grafting process is ideal when a person needs to produce more queen cells than the bees will produce during the swarm season. In the history of beekeeping, the use of the transferal method as explained by Doolittle was one of the major developments of beekeeping that provided beekeepers a method of mass-producing queens. Breeder mothers were being imported from Italy and other regions of Europe and in a few decades beekeepers used the grafting method to replace hundreds of queens. They replaced the dark and less productive northern European stocks dating back to the first bees introduced into Virginia and New England colonies with the gentle and more productive yellow Italian race that helped develop modern beekeeping in North America.

Langstroth produced queen cells by dequeening colonies and harvesting the best of the queen cells these colonies produced. In fact, he would keep three colonies in a rotation of periods with and without queens. I suspect many beekeepers would find this to be an excellent method to produce a few hundred queen cells each year, especially if they select for large queen size (and we discuss this in our section on queens.

BAN MOST CHEMICALS

No Pesticides or Antibiotics Allowed

No pesticides or antibiotics are allowed, although the use of natural organic acids such as formic and oxalic acid may be used for mite control

Advantages	Disadvantages
Chemical contamination is reduced in colonies.	Faced with colony loss, chemicals can be the answer.
Potential of less chemical contamination.	Formic and oxalic acids can also damage colonies.
Less damage to queens and drones (and sperm).	Natural selection will take several years or decades.
Natural selection eventually provides the answer.	

With the discovery of varroa mites in the United States, a persistent group of beekeepers and researchers criticized the beekeeping industry's quick adaptation of chemical treatment. I joined that group early on, feeling that we would never find a magic chemical or 'silver bullet' that would eliminate or mitigate the impact of the parasites. Also, I hoped a genetic answer could be found.

Of course, large chemical companies screened their existing miticides to find known mite-killing chemicals that could be tested for varroa control. Some were effective killing varroa mites, but the consequences of their long-term use were poorly understood. From a global perspective, the market for a popular miticide compound is potentially very profitable.

University and government research programs want fast results. Researchers working on non-chemical treatments of any pest have relatively small budgets and limited support from the beekeeping industry. It has taken many years for their work to be done, tested and accepted. It has taken over 30 years for this to happen with chemical free methods of varroa mite control. The use of chemical-free mite treatments and genetic selection offer the beekeeping industry the option of having vigorous colonies without mites.

Instead, varroa-killing chemicals were released, the mites were controlled, and then the mites developed resistance to the chemicals, all in the matter of a very few years. Unfortunately, managed bee colonies that were chemically treated could not develop resistance, or tolerance, to the mites because natural selection was not allowed to work.[63] Isolated feral (unmanaged by humans) colonies, growing remotely in bee trees and other spaces, have developed natural protection against mites.[64]

Mite tolerant stocks were selected by university and government researchers, often on limited budgets. Russian, Suppressed Mite Reproduction, Minnesota Hygienic, Varroa Sensitive Hygienic and High Grooming (MBB) stocks have been developed. Of note, all used grafting and instrumental insemination technology. We have the tools, the genetics exist, but the beekeeping industry has lacked the motivation, nerve and financial reserves to go chemical free. Certain stocks may be difficult to obtain, requiring beekeepers to plan ahead and plan service from queen producers.

Improved queens cost a bit more than general run queens, more than a chemical treatment, so the industry has opted for treatment. It is possible to attend a bee school or course and never hear a word about mite tolerance or resistance mentioned during the entire day. Discussing only chemical treatments is just wrong.

The result has been the widespread contamination of the beeswax in North America, the development of resistance to certain chemicals, and the stifling of any evolution and selection of natural controls of mites. Those mechanisms are known, have been used in certain parts of the world, but are not popular with U.S. and Canadian beekeepers.

Antibiotic use by beekeepers now requires approval by a veterinarian. This should limit the contamination of the human food chain, taking them out of honey and other products.

Final Thoughts

The general concept of letting bees do what they have done for hundreds of thousands of years should be promoted. If the aim of biodynamic beekeeping is to allow bees to develop according to their nature and to reduce stress (on the bees), I agree that many of these biodynamic principles fit that objective.

I understand and admit that, too often, beekeepers look at honey bees as either pampered pets or cattle in a feed lot that require extensive manipulation with total disregard to their biology, behavior or natural instincts. Some critics point their fingers at commercial, for-profit beekeepers and say that what they do is wrong and that they abuse the bees. Yet I have worked with commercial beekeepers who skilfully run their bees with tremendous regard for the bees and respect for their nature. I wish this regard was universal.

At the same time, I often see new and small-scale beekeepers employ mis-management and disrespect toward the bees. They are well-intentioned but unfortunately poorly informed and completely mis-trained beekeepers. Hobby beekeepers kill bees by loving them to death, delaying colony growth and negatively impacting the hive's ability to survive.

We all have so much to learn to keep our hives alive and minimizing their stress. And our stress too.

8. NOTES

Numbers correspond with the footnote numbers in the text of the book.

1. Before tracheal and varroa mites were found in the USA, beekeepers experienced and expected losses of below 15 % on an annual basis.

2. Fire is totally effective for American foulbrood but fire use is restricted in some areas. Irradiation kills pathogens and parasites on the equipment. While expensive, it is usually less expensive than replacing with new equipment.

3. Egg-laying rate is an essential reflection of how much food the queen is being fed, which is influenced by weather conditions and the available supply of pollen and nectar. This includes stored food. The bees control queen feeding, influenced by their genetics—certain races of bees are quick to buildup in the spring, while other races wait until conditions are just right. Then these colonies increase their population very rapidly.

4. beeinformed.org. The acceptable winter loss rate, shown in the grey bars, is the average percentage of acceptable winter loss declared by the survey participants each year of the survey. By way of explanation, for 2019 the Preliminary Total losses were 40.7% The overwinter loss rate of 37.7% is the highest in 13 years of sampling when compared to the 12-year average of 28.8% By group: backyard beekeepers experienced 38.8% winter loss; sideliners 36.5% and 37.5% for commercial beekeepers. This suggests that the size of the beekeeping operation does not predict survival success. Thanks to Dr. Dewey Caron for these data. The mail address for The Bee Informed Partnership is: 4291 Fieldhouse Drive, 4112 Plant Sciences Building, College Park, Maryland 20742.

5. Data drawn from the California Almond Board (almonds.com) and related websites.

6. Honey is sold in two markets: The international commodity market sells honey to commercial buyers, cereal companies, bakeries, and breweries. They expect to pay the global market price of honey, which may be $1.50 to $3.00 per pound. As I write this, the average price of light honey in 55 gal. drums is $2.26/lb., while the average price for amber honey is $2.13/lb. The second market is the local honey market. Here, small scale beekeepers sell honey to their local buyers. While I ask $10/lb. for any honey I might sell, others sell for less and a few for more. Fireweed honey, for example, is sold by Alaskan beekeepers to tourists and mead makers for about $25/lb. That is $900 for a 60 pound pail of honey.

7. Dr. Tyler Andre and Charlotte Hubbard are removing combs from the side of a house and fitting them into empty frames. String and rubber bands are often used. Bees eventually attach the frames to the sides with more wax.

8. In the early 1980s I enrolled in a class called "Management By Objective". While intended to lead employees, it helped me to stay on track and reach some of my goals. "Management by Objectives (MBO), also known as management by results (MBR), was first popularized by Peter Drucker in his 1954 book *The Practice of Management*. Management by objectives is the process of defining specific objectives within an organization that management can convey to organization members, then deciding on how to achieve each objective in sequence. This process allows managers to take work that needs to be done one step at a time to allow for a calm, yet productive work environment (Wikipedia).

9. Connor, L. *Bee-Sentials: A Field Guide.* 2012. Wicwas Press.

10. Why Three? In art, three is an artistically balanced unit, as many things are divided into thirds. It takes at least three legs to support a chair or table. "The rule of three perplexes me" was a favorite saying of my predecessor at The Ohio State University, William A. Stephens.

11. In the early 80s I did a series of video tapes on basic beekeeping. I look at them now with their poor sound quality, rough cuts and dated recommendations and am reminded how temporary and dated things become on video. I found out in a hurry that everyone has a Ph.D. in watching TV.

12. The realities of commercial agriculture and commercial pollination services force the use of two-ton trucks and semi-trucks for delivery of colonies for pollination, often hives just removed from almond orchards of California, as well as general colony movement. Smaller trucks and forklifts are used to position the hives around the crop. Wet weather and flooded fields are a huge impediment to these moves.

13. Connor, L. *Increase Essentials.* 2006. Wicwas Press.

14. In some far northern regions beekeepers kill off the bees in the fall because it is cost effective to buy new packages rather than feed bees through the winter. Like all economic debates, there is passion on both sides.

15. In the U.S., land surveyed under the Public Land Survey System (PLSS), a section is an area nominally one square mile (2.6 square kilometers), containing 640 acres (260 hectares), with 36 sections making up one survey township on a rectangular grid containing 23,040 acres (9360 hectares).

16. *Datura* is a genus of poisonous night-flowering plants of the family Solonaceae. Also known as moonflower and devil's trumpets. The pollen and nectar are not poisonous to pollinators.

17. Brother Adam.Beekeeping at Buckfast Abbey. 1975.

18. Lau P, Bryant V, Ellis JD, Huang ZY, Sullivan J, Schmehl DR, et al. (2019) Seasonal variation of pollen collected by honey bees (*Apis mellifera*) in developed areas across four regions in the United States. PLoS ONE 14(6): e0217294. doi.org/10.1371/journal.pone.0217294 Provided by Public Library of Science

19. I use the term mite tolerance rather than mite resistant as I do not feel we have any bee stock that is able to resist the varroa mites at the 100% of the time. Tolerance reflects a balancing act, a relationship, between the host and its parasite.

20. G.H. Cale Jr. Personal communication. 1976-1977

21. A dark-race hybrid, the Midnite Hybrid, was produced as well, with a primary focus on gentleness for small-scale beekeepers.

22. Based on: Thomas Seeley. *The Lives of Bees: The Untold Story of the Honey Bee in the Wild.* 2019. Princeton, 105. I have added my observations on current hive arrangements. Seeley poses a great question: "This means that, on average, a wild colony's nest cavity has only one-quarter to one-half of the living space of a typical beekeeper's hive. At this point we wondered, do wild colonies prefer rather small and snug nesting sites."

23. When queens are produced by letting the bees produce their own queen, the process is called Walk-Away Splits by some beekeepers. The beekeeper ended up with new queens (or at least the strongest queen) and the brood and worker bees from two laying queens. This produced a larger honey crop than the single queen colony would produce. Some beekeepers call this a modified two-queen system. One article I wrote while at Ohio State was reprinted in *Swarm Essentials* by Repasky and Connor.

24. Laidlaw, Harry, Jr. and Page, Robert, Jr. *Queen Rearing and Bee Breeding.* Wicwas Press. 1997.

25. Sammy Ramsey. Personal communication. 2018.

26a. Saskatraz hybrids were scored for brood pattern, chalkbrood, temperament, pollen placement, queen status, queen mark, phoretic mite infestation (MPH), percent brood infestation (worker and drone), and hygienic behavior (freeze test).

26b."According to new research in the Entomological Society of America's Journal of Economic Entomology, the fungicide iprodione, when used alone or in combination with other common fungicides, leads to a significant reduction in the 10-day survival rate of forager honey bees (*Apis mellifera*) when they are exposed at rates common to usage in the field...Given that these fungicides may be applied when honey bees are present in almond orchards, our findings suggest that bees may face significant danger from chemical applications even when responsibly applied," says Juliana Rangel, Ph.D., assistant professor of apiculture in the Department of Entomology at Texas A&M University, and a co-author of the study."

27. From Ellis, J., University of Florida and A. Ellis, Florida Dept. of Agriculture and Commercial Services. FDACS.DPI|EDIS. Accessed online 9 Aug. 2015.

28. Conlon, B.H., A. Aurori, A-I. Giurgiu, J. Kefuss. 2019. A gene for resistance to the Varroa mite (Acari) in honey bee (*Apis mellifera*) pupae. Molecular Ecology.

29. Seeley,Thomas. *Following the Wild Bees: The Craft and Science of Bee Hunting.* Princeton. 2016.

30. We were able to increase drone production during the winter in Florida by using artificial lighting, but this work needs to be repeated to remove any secondary benefit from the heat produced by the lights.

31. The acid content of honey helps restrict fungal development in honey with 18% or less moisture. Honey with higher moisture will experience fermentation in the comb or in containers.

32. Broodlessness causes a brake in brood rearing and also a break in varroa mite reproduction. For more details, see note 41.

33. I routinely use frames from several colonies to establish a new hive, knowing that all the colonies I use are disease free. There cannot be any American foulbrood in any of my colonies as that would shut down my increase making and force me to go into triage mode to prevent the spread of the disease. Also, I do not use frames containing any obvious signs of chalkbrood, sacbrood, or European foulbrood. With viruses, I avoid heavy infestations of: deformed wing, K-wing, crawling workers and black queen cells.

34. Since Guzman made that statement, Ontario experienced high bee mortality from insecticide-treated seed corn, but that does not diminish the threat from varroa mites.

35. Connor, L.J., Varroa Control: Past and Future. The *American Bee Journal*. September 2015.

36. Egyptian study, Acad. J. Entomol. 8(4):174-182, 2015

37. Personal communication with Tibor Szabo, Sr. and Tibby Szabo Jr. (Ontario). They showed me their system of weighing each queen soon after emergence in vials. Queen cells were placed into the vials just before emergence. Once emerged, the young queens were weighed on a very accurate scientific scale. Queens below a certain weight were discarded, and only heavier queens were put into mating nuclei.

38. Deborah A. Delaney, Jennifer J. Keller, Joel R. Caren, David R. Tarpy. The physical, insemination, and reproductive quality of honey bee queens *(Apis mellifera* L.) Department of Entomology, Campus Box 7613, North Carolina State University, Raleigh NC 27695-7613, USA.

39a. Three honey bee subspecies; *Apis mellifera carnica, A. m. ligustica* were used as local imported races and *A. m. lamarckii* categorized as a native race. For each subspecies of honeybee, a total of nine colonies (established in ten-framed Langstroth hive) three for each race were used as mother colonies for queen rearing and supply the mating nuclei with young workers and capped brood.

39b. Some believe comb orientation is determined by magnetic, geologic, prior nest site orientation, psychic or other factors. Only sketchy research supports any of these theories.

40. Koeniger et al. *Mating Biology of the Honey Bee (Apis mellifera)*. Wicwas Press. 2014.

41. Varroa mites enter worker and drone brood just before the hive bees seal the larvae into the cell and the larvae pupate. By creating a break in egg-laying, there is a period of time, starting 9 days later, when the mites have no cells to enter. While they may continue to feed on fat bodies of nurse bees, these bees are getting older too and are less attractive to the mites. While a break in the brood cycle does not eliminate varroa mites, it reduces their ability to reproduce. The "no-egg" interval is short when a mated queen is installed and longer for virgin queens, queen cells and 48-hr queen cells. It interval is longest for walk-away splits.

42. Dewey Caron, Honey Bee Health Coalition 2019. Fluvalinate is not recommended for use because varroa mites are resistant to fluvalinate in most of North America. Before using either Apistan® (fluvalinate) or CheckMite+® (coumaphos), test for resistance via the Pettis test, and monitor colonies and mite levels closely. Mite resistance to synthetic chemicals can be avoided by

using the product according to its label instructions, diversifying varroa mite treatments, and following an IPM strategy.

44. "*Melaleuca quinquenervia*, Melaleuca is a large evergreen tree typically 65 feet in height with a brownish white, many-layered papery bark. Melaleuca trees have extensively invaded South Florida, displacing native vegetation in wetland and upland environments. Native to Australia and Malaysia, melaleuca was introduced into Florida in 1906 as a potential commercial timber and later extensively sold as a landscape ornamental tree and windbreak. It was also planted to dry up the Everglades to decrease mosquito populations and allow for development. Melaleuca forms dense stands resulting in the almost total displacement of native plants that are important to wildlife. In the Everglades, melaleuca trees form nearly monospecific forests in formerly treeless sawgrass marshes, disrupting historical water flows. Melaleuca forests represent a serious fire hazard to surrounding developed areas because of the oils contained within the leaves that create hot crown fires." *Florida Fish and Wildlife Conservation Commission.*

45. About 3/8 of an inch: the functional space of two bees on opposite frames so they do not rub against each other. At 1/4 inch or less, the bees fill the gap with propolis (plant resins and beeswax). When the space is larger than 1/2 inch, they will build a piece of comb. Mind the gap. The bees do.

46. 1972-1976.

47. Genetic Systems Incorporated, Labelle, Florida, established 1976. Established by Dadant and Sons, Harvey York (York Bee Company) and commercial queen producers.

48. https://honeybeesuite.com/what-is-biodynamic-beekeeping/

49. There is some evidence that queens and drones of certain races mate at specific mating altitudes.

50. Perhaps plastic combs can be detoxified by burying them in the ground or in a composting system that will "digest' the wax along with all the environmental pollutants found in modern beehives, allowing biotic agents to destroy bacteria, viruses and other risks to bee colonies.

51. Personal communications with Kim Flottum, Editor, *Bee Culture.*

52. Personal communications with the museum docent at Old Sturbridge Village, Sturbridge, Massachusetts.

53. Lau P, Bryant V, Ellis JD, Huang ZY, Sullivan J, Schmehl DR (2019). Seasonal variation of pollen collected by honey bees (*Apis mellifera*) in developed

areas across four regions in the United States. PLoS ONE 14(6): e0217294. doi.org/10.1371/journal.pone.0217294 Honey bee (*Apis mellifera*) colonies require a diversity of protein-rich pollen in order to rear healthy brood and ensure colony survival. During certain seasons, insufficient or poor-quality pollen can limit brood nutrition. In this study, the authors investigated the variation in pollen collected by honey bees across developed landscapes in California, Michigan, Florida, and Texas over the seasons of the year. The authors tracked a total of 394 sites with at least two hives each in urban and suburban locations. They placed a pollen trap at each hive entrance, which passively collected pollen from foraging bees, and sampled pollen from the traps in multiple months of 2014 and 2015. The researchers used a light microscope to identify pollen grains to the family, genus, and species level where possible. The total overall pollen species diversity varied significantly across all four states, with highest diversity in California and lowest diversity in Texas. Nationally, the total pollen diversity was significantly higher in the spring across all locations as compared to other seasons. These date might help urban planners and gardeners choose plants that can provide appropriate pollen resources to honey bees in developed areas year-round, and plan pesticide treatment regimens around honey bee foraging schedules. The authors add: "This study gives us a comprehensive look at some of the most important plants for honey bees in developed areas, and serves as a foundation for studies related to honey bee nutritional ecology."

54. In the Florida bee breeding program, we started with hives that had very low pollen reserves, and experienced record cold weather while trying to produce drones. I purchased a large quantity of naturally-collected bee pollen from a western state producer. The product appeared to be perfect. Unfortunately, after feeding it to our drone production colonies, we unintentionally discovered the best way to inoculate a colony with chalk brood! Most of the drones we produced in the first brood cycle were killed by the fungus, and our production schedule took a huge hit financially. Looking under lower power microscope we found tiny bits of chalk brood mummies the bees had chewed up. I have not fed bee pollen since.

55. Repasky, Steve and L.J. Connor. *Swarming Essentials: Ecology, Management, Sustainability*. Wicwas Press. 2014.

56. White, E.B. "Song of the queen bee". *The New Yorker*, Dec. 15, 1945.

57. Two or three pounds of bees removed from a hive and shipped with a

freshly mated but unrelated queen to beekeepers hundreds or thousands of miles away from the source.

58. Doolittle, G. *A Year in the Out-Apiary.* 1908, and Connor, L. *Increase Essentials.* 2006.

59. Do not to shake a queen cell at any point in its development.

60. Production queens are those used in a for general beekeeping purposes, honey production and brood rearing. In the production of queens, you need many strong colonies to support the breeder queens and the hives staring and finishing queen cells.

61. We can improve on this if we adapt Brother Adam's breeding isolation plan used to develop the Buckfast bee. Unique crosses were made in an isolated location by moving in drone colonies of a target stock, and then using another line of queen cells in the mating nuclei. This allowed the mating of several crosses during one season. That way we can reduce the number of islands (remote mating yards) we need for isolation.

62. There is some evidence that queens and drones of certain races mate at unique mating altitudes.

63. The beekeeping industry has rejected to consider a genetic answer in the past. When offered stock resistance to American foulbrood in the early 1950s by Dr. G.H. "Bud' Cale, Jr. as a potential feature of the Starline Hybrid bee, queen breeders and large commercial beekeepers opted instead to use antibiotics as treatments to control the spore-forming bacillus we continue to see in our hives over half a century later.

64. Personal communication, Dwight Wells. Dwight has been working with several groups to develop queen lines from areas of the Midwest that have not had intensive chemical treatment or contamination with by commercial beekeepers. Due to his work, in Pennsylvania, a SARE grant looked at the mite-biting behavior. In the final report: *Assessment of Project Approach and Areas of Further Study:* "The study did get a good sampling of feral bees to look at and evaluate, so we successfully performed our experiment. Many of the swarms were really awful bees, mean, aggressive, disease ridden, non-productive; but, a few were exceptional bees and show promise. We know that we have any real 'winners' from our feral swarms that we captured, and continued maintain and monitor them."

INDEX

reproduction by swarming 101
Rhus aromatic 158
Rigney, J. 144
Rinderer, Thomas 68
ripe honey 98, 104
robbing 90, 170, 195, 197
Rothenbuhler, Walter 10, 155, 200, 203
Roundup 160, 162
royal jelly 23, 96, 104, 133, 142, 154, 182, 193, 205
Rueter, Gary 30
Russian Bee Breeders 65
Russian bees 65, 67-68, 119-120, 131, 208
Sammataro, Diana 27
San Francisco Bee Club 34
Saskatraz 69, 169
scenting bees 114
screened bottom boards 128
sealed brood 70
sealed honey 77
second or secondary swarm 86, 101
Seeley, Thomas 27, 89
sex determination 112, 203
sexual reproduction 96, 100
sexually mature drone 96-97, 202
shaking bees 82-83, 87, 130-131, 194-195, 198
shaking jar 131
shorter development 72, 119
skeps 173, 185
slice of bio 174-175
slumgum 183
small hive beetles 11, 17, 42, 191
small queens (see queen size)
smaller bees 179-180
smaller colonies 67, 145
Smith, Jay 205
Smith, P. 71

Smith, Todd 95
snow machines 44
specialized anatomy 106-108
sperm 84, 98, 100-101, 107, 120, 143, 149, 162, 184, 201, 204, 207
spermatheca 98-101, 148-149, 201
Spivak, Marla 70
spotted knapweed 53, 157
spotty brood 151
staggered planting 37
Starline hybrid 61-65, 70, 120, 139-140, 180, 202
starting beekeeping 20, 23, 29, 54, 75, 120-121, 195
Sterk, Bo 185-186
sting structure 110-111
stinging behavior 35, 61, 65, 90, 114
stings 35, 61, 65, 90, 110-111, 114
strains selected for tolerance 57, 66, 204
sub-lethal damage 45
sub-species 102, 180
sucrose 104, 109
sumac 158-159
summer losses 132
sun 42
Sunbelt queens 80, 198
sunflowers 155
super sisters 100
supersedure 84, 96, 123, 143-144, 146, 152, 198-200, 204, 206
support hives 22, 79
support network 35
suppressed mite reproduction 58
surfactants 105
survival challenge of swarming 87
survives burning 159
survivor stocks 67-69, 72, 90, 119, 128
sustainable 30, 46, 53, 57, 77, 87, 89, 93,

116-118, 124, 159, 170, 184
swarm (swarming) 47, 67, 72, 75, 78-79, 82, 86-89, 91-92, 94, 96, 101-102, 105, 113, 122-123, 143, 153, 155, 157, 164-165, 178-179, 185, 192-197,200, 204, 206
swarm capture 86-88, 91, 194
swarm cells 79, 102, 195
Swarm Essentials 2
swarm season 101-102, 206
synergistic effects 45
synthetic miticides 161
Tarpy, David 64
teachers 32, 34, 36, 53, 175
temperature 48, 95, 100, 102, 154
Tew, Jim 27
thermal imaging 95
three-frame nuclei 39
Tilia spp. 158-159
tracheal mites 65, 119, 133, 162-163
tracheal tubes 101, 148
tractors 44
trailer 37-39, 166
transportation 177
treated lawns 45, 53
triangular relationship 60, 119
tropics 43, 95
Tropilaelaps spp. 74
truck stuck 40
trucks 18, 38, 44-45, 122, 145
two queens in same hive 64, 149
ultraviolet 108
University of California 31
University of Florida 30
University of Guelph 29-30, 33, 124
University of Minnesota 30, 70, 187